Battery-less NFC Sensors for the Internet of Things

Battery-less NFC Sensors for the Internet of Things

Martí Boada
Antonio Lazaro
David Girbau
Ramón Villarino

ISTE

WILEY

First published 2022 in Great Britain and the United States by ISTE Ltd and John Wiley & Sons, Inc.

ISTE Ltd
27-37 St George's Road
London SW19 4EU
UK

www.iste.co.uk

John Wiley & Sons, Inc.
111 River Street
Hoboken, NJ 07030
USA

www.wiley.com

Library of Congress Control Number: 2022939222

British Library Cataloguing-in-Publication Data
A CIP record for this book is available from the British Library
ISBN 978-1-78630-836-8

Contents

Introduction

I.1. From barcodes to the Internet of things

Streamline industrial processes have always been a priority to reduce time and, therefore, to reduce costs. The identification of goods is an important factor to consider in almost any process since it is necessary to track and get specific information about them, hence the importance of automation. The most common method is to use barcode systems on a printed label with an identification number, which is read by an optical device connected to a computer (Palmer 2007) (see Figure I.1)

Figure I.1. *Barcode reading. For a color version of this figure, see www.iste.co.uk/girbau/sensors.zip*

Barcodes have been included in several places of work such as libraries (Singh and Sharma 2015). They have medical uses, like in identifying blood transfusions (Murphy and Kay 2004) or in the administration of medication (Macias *et al.* 2018), proving that it is an easy, fast and accurate automatic collection method. Linear barcodes are the most common technology related to automatic identification data capture (AIDC). Usually, the reading methods use both hardware and software implementations. However, there are fully implemented methods with hardware (De Maeyer *et al.* 2003) and software (Muniz *et al.* 1999) solutions. Nonetheless, barcodes only contain information in one dimension (1D), containing therefore a limited amount of information. To enlarge the data representation capability, two-dimensional (2D) codes have been developed. Within the collection of 2D barcode symbologies (e.g. MaxiCode, PDF 417, codablock F, Aztec), two of them are the most widespread. The first choice is the data matrix, standardized by the ISO/IEC 16022 (ISO/IEC 2000a), and the second one is the quick response (QR) codes, standardized by the ISO/IEC 18004/2000 (ISO/IEC 2000b). Data matrix capacity is 2,334 alphanumeric characters, or 1,556 ASCII characters, and it uses between 30% and 60% less space than a QR code (Semacode 2006) (see Figure I.2). However, it does not allow us to encode Japanese characters, which is the main reason why QR codes have been more extended worldwide.

a) b)

Figure I.2. *"http://google.com encoded" as (a) a QR code and (b) a data matrix*

The implementation of 2D barcodes in the industry brings advantages to some, such as increased data capacity or the replacement of special-purpose laser scanners with digital cameras, which comes with a cost reduction and improves image processing by using dedicated software and algorithms (Ohbuchi *et al.* 2004; Falas and Kashani 2007; Liao and Lee 2010). Nowadays, smartphones are the prevailing devices with a built-in digital camera. Furthermore, the computational capability of the smartphone allows the installation of applications that read all kinds of barcode symbologies. Since QR codes are supposed to be widely used by any type of smartphone simply by taking a photo of the code, light conditions and possible damages or other alterations on the printed code must be considered (Liu *et al.* 2008) as well as security issues (Kieseberg *et al.* 2010).

Within AIDC systems, radio-frequency identification (RFID) (Finkenzeller 2010) arrived to overcome barcode limitations. An RFID system consists of two differentiated parts: a reader (also called an interrogator) and a tag (also known as a transponder). The reader generates a signal which is received by the tag. This signal is modulated at the tag embedding its internal information (e.g. an ID), and then it is sent back to the reader to be interpreted. Table I.1 lists some of the differences between barcodes and RFID.

Barcode	RFID
Line of sight	No need of line of sight
Read one by one	Multiple reads simultaneously
Fixed content	Updatable content
Manual tracking	Can be automatically tracked

Table I.1. *Comparison of barcode and RFID*

The spread of RFID technology at large scale began during the first years of the 21st century, when three major organizations (Walmart, Tesco and the US Department of Defense) opted to use this technology, streamlining the tracking of stock, sales and orders (Want 2006). The expansion of RFID came along due to the need to analyze and adapt its use for commercial applications and the need for regulations (Kelly and Erickson 2005) as well as for integrating methods for security and privacy (Juels 2006). RFID have been widely used for all kinds of applications (Weinstein 2005) such as object localization (Hahnel *et al.* 2004; Ni *et al.* 2014), car parking (Pala and Inanç 2007), food traceability (Abad *et al.* 2009), logistics services (Chow *et al.* 2006) and navigation assistance for blind people (Chumkamon *et al.* 2008). The interrogators used to read the tags can be connected to the Internet, and are thus able to send the collected data of the identified objects to the cloud in real time, which leads to the so-called Internet of things (IoT) (Jia *et al.* 2012).

Passive RFID systems reduce the price of the tags to a few cents as well as to expand lifetime due to the lack of batteries. However, there is still an important drawback: the price of readers, besides the fact that each reader works in a specific frequency; thus, they are only useful for certain systems. Nevertheless, today, there is a subset of protocols within RFID technology, which is growing fast, known as near-field communication (NFC), which is embedded in most smartphones sold today due to its worldwide implementation as a payment mechanism (Pal *et al.*

2015; Ramos-de-Luna *et al.* 2015). Having NFC-enabled smartphones brings new opportunities to the usage of RFID applications, since the reader is now in the pocket of millions of end-users and almost always has an Internet connection. This makes NFC technology a suitable option in the IoT paradigm to be used as a bridge between *things* and the *Internet* without the need of having all the *things* directly connected to the cloud (Cerruela García *et al.* 2016; Shah and Yaqoob 2016).

I.2. RFID technology

The first patent for an active RFID transponder capable of rewriting stored data was created by Mario W. Cardullo in 1973 (Cardullo 1973). Since then, many types of RFID have appeared. RFID systems can be classified according to features such as the operational frequency, the range, the physical coupling method or the need for batteries in the transponder (i.e. passive/active systems). Table I.2 shows the classification of RFID based on the three sub-1GHz different standardized operational frequency bands along with their main features.

	LF	HF	UHF
Coupling mode	Inductive (near-field)	Inductive (near-field)	Electro-magnetic backscattered (far-field)
Operating frequency	125–134 kHz	13.56 MHz	860–960 MHz
Regulation	Worldwide harmonized	Worldwide harmonized	Different regulations per country
Standards	ISO14223 ISO18000-2	ISO14443 A/B ISO15693 ISO18092 ISO18000-3 NFC Forum	ISO18000-6 B/C EPC Class 1 Gen 2
Environmental influences	Water: no problem; metal: low attenuation	Water: low attenuation; metal: critical	Water: critical; metal: less critical
Applications	Animal tagging	Product identification/public transport/payment	Container ID/ Logistics/authentication

Table I.2. *RFID technologies. LF = low frequency, HF = high frequency, UHF = ultra high frequency*

An overview of an RFID system is shown in Figure I.3, consisting of a reader and a remote tag, and shows where the signal sent by the reader and the tag response is represented in both time and frequency domains.

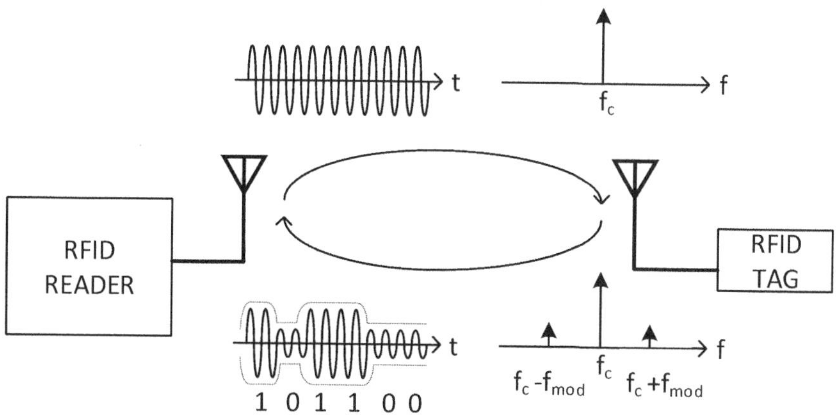

Figure I.3. *RFID system overview*

RFID devices can be divided into two main categories: passive and active. Active tags need to be powered by a battery or by a connection to a power source, whereas passive tags use the energy received from the reader to operate. Active tags have better performance than passive tags, reaching longer ranges and allowing larger memory banks; however, the cost, size and lifetime make them not as interesting as passive tags for the retail trade. Passive systems have a limited range determined by factors such as reader parameters, propagation environment and tag characteristics. Table I.3 summarizes different characteristics of passive and active tags.

	Active	**Passive**
Power source	Internal	Harvested from the RF field
Required signal strength (reader to tag)	Low	High
Available signal strength (tag to reader)	High	Low
Communication range	Long range	Short range
Multi-tag reading	Thousands of tags	Several tags
Sensing capability	Continuously (all data retrieved when the reader approaches)	Only when the tag is within the reader's range
Data storage	Large	Small
Cost	High	Low
Typical operational frequency	433 MHz, 2.45 GHz, 5.8 GHz	128 kHz, 13.56 MHz, 915 MHz, 2.45 GHz

Table I.3. *Active and passive RFID tag comparison*

I.3. Near-field communication (NFC)

NFC is a subset of protocols within RFID technology, which operates at 13.56 MHz, mainly aimed at mobile or handheld devices.

There are three possible operating modes for an NFC tag based on the ISO/IEC 18092 (ISO/IEC 2013), ISO/IEC 14443 (ISO/IEC 2018) and ISO/IEC 15693 (ISO/IEC 2010) standards:

– Card emulation: the device acts like a contactless card. Typically, one uses the device for payments or ticketing. Can be based on ISO/IEC 14443 or 15693.

– Reade/write: the device can read a tag and overwrite the content on it if the tag allows it. It is used for smart posters to get a URL or the sensed data from a tag. It can be based on ISO/IEC 14443 or 15693.

– Peer-to-Peer (P2P): it allows the exchange of data between two devices. It is used to transfer data such as photos. It is based on ISO/IEC 18092.

There is a fourth mode, whose technical specification is still awaiting approval:

– Wireless charging (WLC): the power transfers from one device to another, such as a smartwatch or fitness tracker. Four power transfer classes are specified: 250, 500, 750 and 1000 mW.

Five types of tags are defined by the NFC Forum, each of them with certain characteristics, as shown in Table I.4.

	Type 1	**Type 2**	**Type 3**	**Type 4**	**Type 5**
Technology	NFC-A	NFC-A	NFC-F	NFC-A NFC-B	NFC-V
Standard	ISO14443A	ISO14443A	JIS 6319-4	ISO14443 A/B	ISO15693
Memory size	Up to 2 kBytes	Up to 2 kBytes	Up to 1 MByte	Up to 32 Kbytes	Up to 64 kBytes
Transmission speed (kbits/s)	106	106	212 /424	106	Up to 26.5
Unit price	Low	Low	High	Medium/High	Low/Medium
Range	Proximity	Proximity	Proximity	Proximity	Proximity and vicinity
Chip examples	Topaz	Mifare ultralight	Felica	Mifare DESFire	ICODE SLI
Use case	Business cards, smart poster	Event tickets, transit tickets	e-money, membership cards	Transit tickets	Packaging, ticketing

Table I.4. *Types of tags defined by the NFC Forum*

I.3.1. *State of the art: current NFC applications*

As the IoT scenario grows, the number of devices and the ways people interact with these devices grow with it. Even though NFC has the shortest range among radio frequency technologies, it has been widely extended due to its security, compatibility, friendly user interface, large amount of applications and low-cost components. As already mentioned, the main reason for the current increase in the integration of NFC in almost any smartphone is explained by its usefulness regarding contactless payments, making it increasingly indispensable for the end-user, as happens with other wireless technologies such as Wi-Fi or Bluetooth. Figure I.4 shows the number of NFC-enabled mobile devices worldwide between 2012 and 2018. Between 2013 and the end of 2018, worldwide shipments of NFC-enabled smartphones increased by 325%. Market estimations expect that by 2022 more than 90% of smartphones will be equipped with NFC.

The different standards of the NFC Forum allow us to select between the different standardized technologies to find the most suitable option for each case scenario (Ok *et al.* 2010; Coskun *et al.* 2015). Each operation mode has its own range of applications. The main advantages of each mode are:

– Card emulation: it replaces physical objects, such as mobile payments and access control.

– Read/write: requires a low physical effort (*tap-and-go*). It increases mobility, is adaptable to many scenarios and is easy to implement.

– P2P: easy way to exchange data at low rates.

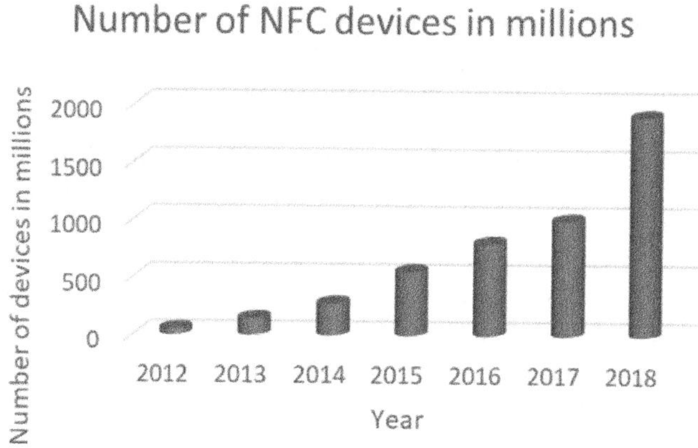

Figure I.4. *Number of NFC-enabled mobile devices worldwide from 2012 to 2018 (in million units)*

Card emulation mode is mainly used to replace physical objects like credit cards (Pasquet *et al.* 2008; Pourghomi *et al.* 2013), tickets (Ghìron *et al.* 2009) and coupons (Ghìron *et al.* 2009). Read/write mode is the most extended mode of operation since it encompasses a huge variety of uses such as smart posters (Ruiz and Gómez-Nieto 2009; Wu *et al.* 2012; Boes *et al.* 2015), healthcare applications (Fontecha *et al.* 2011; Puma *et al.* 2012; Köstinger *et al.* 2013; Sethia *et al.* 2014), medication dispensing (Lahtela *et al.* 2008; Morak *et al.* 2012; Alabdulhafith *et al.* 2013), catering meal services (Siira and Haikio 2007), social networking (Fressancourt *et al.* 2009), museum interactions (Blöckner *et al.* 2009), attendance supervision (Ervasti *et al.* 2009), enterprise workflow control (Müller *et al.* 2009), sales assistance for retailers (Karpischek *et al.* 2009) and to control multimedia players (Sánchez *et al.* 2007). P2P mode has fewer applications since it is mainly focused on the exchange of data. However, there are some application studies for its use in cases where other operation modes are more common. For instance, for mobile payments, P2P mode adds some characteristics like customizable confirmations and the possibility to avoid the use of a secured element (Mainetti *et al.* 2012). Another possible application of P2P is to transfer data between medical devices and the nursing station where the patients' information is collected (Huang and Chang 2013).

More recent works found in the literature use NFC for sensing applications (Lazaro *et al.* 2018; Cao *et al.* 2019; Cui *et al.* 2019). An extensive range of sensing applications can take advantage of NFC, for instance cold-chain temperature control (Zhao *et al.* 2015; Lorite *et al.* 2017), gas sensors (Azzarelli *et al.* 2014; Escobedo *et al.* 2017), soil moisture (Boada *et al.* 2018a) or tire pressure (Kollegger *et al.* 2017).

In addition, healthcare can clearly benefit from the easy sensing methods that NFC provides. Some representative applications are: body pressure and temperature (Vicente *et al.* 2016; De Oliveira Filho and Do Prado Villarroel Zurita 2017; Jeong *et al.* 2017; Han *et al.* 2018), sweat analysis (Rose *et al.* 2014; Koh *et al.* 2016), glucose monitoring (Anabtawi *et al.* 2016; DeHennis *et al.* 2016), ultraviolet radiation (Shi *et al.* 2018), pH level (Douthwaite and Georgiou 2017; Rahimi *et al.* 2017) and electrocardiography (ECG) (Wu *et al.* 2015). The implantable device is a promising challenge and a widely studied topic (Bazaka and Jacob 2012; Tankiewicz *et al.* 2013; Hannan *et al.* 2014; Altawy and Youssef 2016; Anabtawi *et al.* 2016; DeHennis *et al.* 2016; Samineni *et al.* 2017). The simplicity of the *tap-and-go* philosophy alongside the widespread use of smartphones with an Internet connection makes NFC technology a promising technology for telemedicine. Furthermore, developing sensing devices without a battery allows NFC to become a very interesting technology for implantable devices.

I.3.2. *Passive sensing tags*

The most important NFC integrated circuit (IC) manufacturers have recently launched advanced IC with energy-harvesting capabilities (Lazaro *et al.* 2018). These IC convert part of the energy from the magnetic field generated by the NFC reader to provide a voltage output. This rectified output can power external electronic devices such as low-power microcontrollers or sensors. The progressive introduction of these ICs into the market enables the development of low-cost battery-less sensors. These devices can be used in sensing applications that must eventually be read on a user's request, which must work for a long time and in short ranges. Unlike other energy-harvesting technologies, battery-less NFC-based sensors have the advantage that a current up to a few mA can be obtained (with some external circuitry added to the NFC IC) using standard smartphones as readers.

In this context, this book is motivated by the possibilities that NFC recently brings in the IoT scenario to create short-range green (without battery) sensing applications, thus avoiding toxic waste while extending device lifetime. To tackle the design of NFC sensors and its applications, some important issues must be addressed. The main constraint is to ensure, beyond the communication, an efficient and dependable energy transfer. Guaranteeing a reliable power supply is pivotal to develop sensors embedded on passive tags. An extensive, systematic analysis on this issue is done in this book. It must be noted that avoiding batteries opens the door to applications which otherwise would not have been feasible. For instance, when using sensing tags to control food quality, the toxicity of a battery should be dealt with by using expensive coating methods. This also happens to body-implantable devices, which have the same problem with battery toxicity. Furthermore, when developing implantable devices, the size is vitally important. Hence, the benefit of evading the use of batteries.

Another important issue addressed in this book concerns the tag design, the reduction of its dimensions and specifically the proposal of a systematic procedure to design the antenna, tune and adapt it to the IC. Beyond the specific application of the tag, the antenna design becomes an important factor since it must be incorporated into compact devices (i.e. smartphones), surrounded by metallic parts that degrade its performance or can be integrated in severe environments (i.e. liquids, body, etc.). Antenna performance becomes more important when the inductive link transmits energy to power up the tag circuitry. Several works have been devoted to improve the performance of the antenna, most of them using ferrite sheets with high relative permeability to avoid the effect of eddy currents. Lee *et al.* (2014) propose to add a parasitic loop on the outer side of a conventional loop antenna, and the use of a ferrite–polymer composite sheet, which is less expensive than ferrite sheets. In Chen and Zhao (2016) and Zhao and Chen (2016), an

"8"-shaped antenna as an improvement of the Murata ZB antenna (WO2013115147A1 2013) is presented to overcome the shielding effect of the metal cover. A five-turn coil design for tablet PC devices is introduced in Chung and Yang (2016), taking into account the fact these devices used to have large touch panels and narrow borders, hence the interest in developing rectangular antennas to fit in the frame space. The proposed antenna size is 41.5×7.5 mm, with ferrite sheets on both sides. Another solution for metal-case devices with non-uniform meandering lines and partial coverage of ferrite is presented in Zhu *et al.* (2017), proposing a six-turn coil with a six-sided irregular hexagonal shape.

A variety of passive sensing NFC tags can be found in the literature. The simplest ones consist of an ON/OFF sensing. Within this category, different methodologies are observed. For instance, in Sidén *et al.* (2011), a tag composed of an SL13A (AMS 2014) chip to detect the presence of urine in a diaper is based on measuring the resistance between two parallel conductors, allowing it to detect if the diaper is wet or dry when the conductors are short-circuited by the urine. An ON/OFF temperature indicator for supply chain monitoring system is presented in Lorite *et al.* (2017). It consists of a critical temperature indicator (CTI) based on a solvent melting point. The tag, based on the Melexis MLX90129 (MELEXIS 2012) as an NFC interface, measures the change in resistance of multi-walled carbon nanotubes (MWCNT) connected to a couple of copper wires. Another methodology is used in Ma *et al.* (2018) for a food spoilage sensor. This sensor relies on the detection of the total volatile basic nitrogen (TVBN) by placing nano-structured polyaniline (PAni) between the NFC coil terminals, taking advantage of its change on resistance when exposed to TVBN, inducing a short-circuit when the TVBN is above a threshold, thus modifying the antenna impedance and making the tag impossible to read. This approach can lead to a false negative reading, since when the tag cannot be read, the user must interpret that the food is spoiled, but it could be that the impossibility to read it comes from another issue. The same issue may occur in Yambem *et al.* (2008), where a passive LC resonating sensor tag for detecting humidity in diapers is shown. In this case, the resonance frequency (12 MHz) is shifted by the presence of urine near the antenna, thus making it impossible to detect the tag by the non-standardized custom reader used.

Other works use ICs that have an NFC interface as well as an embedded sensor. Kifle *et al.* (2019) present a tag consisting of a square 25×25 mm printed coil on a flexible biocompatible plastic substrate, connected to a temperature logger with an embedded NFC interface (NHS3100) (NXP Semiconductors 2018c) from NXP. In Jeong *et al.* (2017), a tag based on the RF430FRL152H IC (Texas Instruments 2014a) is presented. The tag is designed on a flexible substrate, with an overall size of 74×50 mm. A serpentine-shaped coil is used as an antenna to provide stretchability and flexibility, since this tag is supposed to be adhered to the skin to measure temperature (the sensor is embedded into the same NFC chip), and the ambient light, captured by a phototransistor. The designed circuit is carved onto a copper foil, which is laminated on a thermal release tape. SL13A IC (AMS 2014) also integrates a temperature sensor, and is used in several works (Sidén *et al.* 2011; Escobedo *et al.* 2017; Rahimi *et al.* 2017; Han *et al.* 2018). In Han *et al.* (2018), an array of 65 NFC sensors for pressure and temperature are placed on a bed, for mapping these parameters in specific areas of the body, creating a wireless sensor network. A dual-antenna system between the mattress and the body provides the energy and gets the information from the sensors.

In order to extend the functionality and capability of the NFC sensing tags, some works incorporate a microcontroller (MCU). Since the MCU has to be energized from the rectification of the RF received signal, it is crucial to use low-power chips. In Zhao *et al.* (2015), an open source programmable sensing and computationally enhanced platform named NFC-WISP (wireless identification and sensing platform) is presented. This work is based on a previous study of an RFID-based battery-less programmable sensing platform (Sample *et al.* 2008), and it consists of a printed circuit board with a 42×53 mm antenna and a software defined NFC tag (ISO/IEC 14443 protocol) on a microcontroller, TI MSP430F5310 (Texas Instruments 2018). This device incorporates an E-ink screen, and the conversion from RF to DC is done with an external full-wave rectifier. It incorporates a temperature sensor as well as an accelerometer. The NFC-WISP is used as a basis in De Oliveira Filho and Do Prado Villarroel Zurita (2017) for the non-invasive temperature sensing of premature newborns in neonatal incubators, placing the tag on the incubators' mattress. Escobedo *et al.* (2017) introduce an NFC tag composed of the SL13A NFC IC and a PIC16LF1703 (Microchip Technology 2015) microcontroller for gas concentration determination. It contains four chemical sensors with optical response for measuring oxygen, carbon dioxide, ammonia and relative humidity surrounding a white LED light on them. It also uses the temperature sensor embedded in the SL13A. A recent work presents an electronic platform for general-purpose sensing (Escobedo *et al.* 2019). The system allows voltammetry, potentiometry, amperometry and electrochemiluminescence (ECL) analysis based on a screen-printed electrode. The design has been applied to glucose, pH and H_20_2 determination. The system uses a PIC16LF1703 microcontroller connected to an AS3955 (AMS 2017) as an NFC interface.

As mentioned above, NFC passive tags are also very interesting for medical applications due to the lack of battery and the possibility of developing implantable medical devices (IMD). A miniaturized flexible tag is presented in Kim *et al.* (2015). Even though it says that it can be used for biosensors and electronic implants, and two chips with energy-harvesting capability are used (M24LR04E (ST Microelectronics 2017b) and NTAG216 (NXP Semiconductors 2015b)), the study focuses on the tag's antenna miniaturization, developing a 7×5 mm tag, but does not implement the sensors. It analyzes the effects on the electromagnetic characteristics when the tag is bent. Wikner *et al.* (2017) propose a study of planar coils printed on a flexible substrate with a different number of turns, conductor widths and other physical parameters for implantable temperature sensors using the SL13A NFC IC. The measurements have been evaluated by implanting the tag into a porcine model able to communicate with a commercial NFC-enabled smartphone with depths of up to 10 mm. The communication between the tag and reader is done by using a Texas Instruments evaluation board which includes a software to measure the received signal strength indicator (RSSI).

Most of the works regarding implantable battery-less devices found in the literature use application-specific integrated circuits (ASIC), which are not standardized and need specific readers to supply the energy and retrieve the data from the tag. An ASIC is, for instance, used in Anabtawi *et al.* (2016), presenting a battery-assisted, implantable, passive, amperometric, electrochemical glucose sensor with an off-chip printable battery, which stores the harvested energy, allowing for continuous monitoring. In DeHennis *et al.* (2016), the authors present a glucose-implantable sensor based on an ASIC that implements the ISO/IEC 15693. It consists of an LED whose light is reflected on a biosensing fluorophore, and a photodiode for detecting the fluorescent-reflected emission.

Beyond the use of the 13.56 MHz ISM (industrial, scientific and medical) band, some works use the same principle but with different frequencies. This is the case of Lin *et al.* (2016), for example, where a battery-less implantable neuro-electronic interface suitable for studying deep brain stimulation (DBS) working at 10 MHz is presented. The implant has eight electrodes which can interact with eight different brain regions bi-directionally. The implant communicates with an external wearable transceiver by means of two coils using LSK (load shift keying) modulation, used for both data and power transfer, reaching transmission speeds up to 2 Mbps. The coil sizes are 20 mm and 30 mm for the implanted one and the external one, respectively. The external module relays the data to a computer or portable device via Bluetooth.

Table I.5 summarizes the passive sensing tags reported in the literature mentioned above, listing the NFC chip used, the sensed parameters, the size and some additional comments for each.

As it can be seen, the type of application where passive sensing tags can be applied is large, and it is expected to keep growing, along with the number of smartphones with NFC capability and the number of NFC ICs.

Ref	Chip	Sensor	Size (mm)	Comments
Lorite *et al.* 2017	MLX90129	Temperature	32×51	ON/OFF
Zhao *et al.* 2015	Software defined in an MCU (MSP430)	Temperature, accelerometer	85×54	Open source platform
Escobedo *et al.* 2017	SL13A	Multigas	65×40	Optical
Kollegger *et al.* 2017	ASIC	Tire pressure	35×15	
Han *et al.* 2018	SL13A	Temperature, pressure	16×16	Sensor network
Jeong *et al.* 2017	RF430FRL152H	Temperature, phototransistor	74×50	Flexible
Anabtawi *et al.* 2016	ASIC	Glucose		
DeHennis *et al.* 2016	ASIC	Glucose	3×15	Implantable
Rahimi *et al.* 2017	SL13A	pH		
Ma *et al.* 2018	-	Nitrogen		ON/OFF
Kifle *et al.* 2019	NHS3100	Temperature	25×25	Flexible
Escobedo *et al.* 2019	AS3955	General-purpose	31×20	
Kim *et al.* 2015	M24LR04E NTAG216	Not implemented	7×5	Flexible, miniaturized
Wikner *et al.* 2017	SL13A	Temperature		Implantable
Lin *et al.* 2016	ASIC	Brain stimulation	20×20	10 MHz

Table I.5. *Sensing tags reported in the literature*

I.4. Scope, objectives and organization of the book

This book focuses on the study of NFC battery-less sensors using commercial smartphones as a reader. To this end, the scope of this work is framed within RFID, in the HF band, using NFC Forum standards, in read/write operational mode with passive tags in the NFC-V technology. To this end, it provides an in-depth study on the most important aspects involved in the design and integration of NFC passive sensors read by means of a smartphone within the growing IoT scenario. The key points addressed in the book are listed below:

– In-depth parametric study of the operation and constraints of a wireless power transfer applied to NFC in real-case scenarios using commercial smartphones as a reader.

– Study of the read range and the energy-harvesting range.

– Study, design and implementation of custom passive tags. Analysis of the behavior of different antenna topologies, and manufacturing materials, as well as comparison of the performance of different NFC IC and different power requirements.

– Demonstration of NFC capabilities and advantages to be used in passive sensing devices, using both analog and digital sensors.

– Several case studies are proposed which show selected sensing applications used as examples to demonstrate the wide possibilities of the technology, where additional concepts are explained, and where results coming from the joint use of NFC with other disciplines is shown.

Chapters 1–5 of the book are organized as follows:

– Chapter 1. Wireless Power Transfer Applied to NFC

This chapter analyzes the architecture, constraints and limitations of a wireless power transfer applied to NFC systems. It describes the theoretical background and main parameters that influence the efficiency of the system, such as antenna design, environmental effects and coupling conditions.

– Chapter 2. Case Study 1: Soil Moisture Sensor

Presents a soil moisture sensor embedded in a battery-less tag able to measure the volumetric water content of the soil as well as the relative humidity and temperature.

– Chapter 3. Case Study 2: Smart Diaper

Describes the application of NFC technology used to detect the level of wetness in smart diapers, designed on a flexible substrate and based on the change in capacitance between electrodes.

– Chapter 4. Case Study 3: NFC Sensor for pH Monitoring

Proposes a prototype for a color sensor for measuring the pH based on color determination of pH-reactive paper-based color strips. The repeatability of the proposed device is addressed as well as a comparison with image processing alternatives.

– Chapter 5. Case Study 4: Fruit Ripeness Sensor

Studies the possibility of using a color sensor to analyze fruit ripeness. An analysis of the different color components of different color spaces is presented, along with the use of a machine learning algorithm used to create a classification matrix used in an Android application to discriminate the ripeness state.

1

Wireless Power Transfer Applied to NFC

1.1. Introduction

Our knowledge of physics determines four different fundamental forces: two of them produce forces at subatomic distances and govern nuclear interactions, and the other two, gravity and electromagnetism, produce long-range effects. Each can be mathematically described as an electromagnetic field. Michael Faraday was the first to describe one of these forces in 1831. He studied the magnetic field generated around a conductor through which a current flows. A few years later, in the 1860s, James Clerk Maxwell developed and unified the theory of electromagnetism, summarizing it in four equations that now bear his name. However, it was Heinrich Hertz who, in 1887, proved the existence of the electromagnetic waves predicted by Maxwell's equations, inventing the dipole oscillator and the resonator (Buchwald 1994). Radio wave evidence was the beginning of wireless power transfer (WPT), and Nikola Tesla was one of the researchers most involved in this regard. He invented radio frequency resonant transformers (Tesla coils) using inductive and capacitive coupling when trying to develop a wireless lighting system. In 1921, he stated: "Power can be, and at no distant date will be, transmitted without wires, for all commercial uses, such as the lighting of homes and the driving of airplanes. I have discovered the essential principles, and it only remains to develop them commercially." (Wisehart 1921)

Despite the efforts and dedication that Tesla put into WPT, it never achieved commercial success due to the dispersion of wireless power, which mainly depends on the operating frequency and the size of the transmitting antenna. Later, in the

---

Sections of this chapter have been taken from a previous publication by the authors of this book. from Boada, M., Lazaro, A., Villarino, R., Girbau, D (2019). Battery-less NFC sensor for pH monitoring. *IEEE Access*, 7, 33226–33239. doi: 10.1109/ACCESS.2019.2904109.

1960s, after knowledge had been acquired of working at microwave frequencies, during World War II, where the energy could be focused into a narrow beam, a new era of WPT began (Brown 1984, 1996).

Figure 1.1. *Nikola Tesla in his laboratory in 1899. Image by Dickenson V. Alley, licensed under CC BY-SA 4.0*

Nowadays, three different methods for WPT using electromagnetic fields are employed: inductive coupling, resonant coupling and far field (Shinohara 2012) (see Table 1.1). These three technologies are similar, and they all follow Maxwell's equations. However, each of them has its own characteristics and design constraints.

	Inductive coupling	**Resonant coupling**	**Far field**
Field	Magnetic field	Resonance (magnetic, electric, EM)	EM
Method	Coil	Resonator	Antenna
Distance	Short	Medium	Short to long
Efficiency	Low to high	High	High
Power	High	High	Low to high
Uses	Felica cards, Qi standard (mobile phone charging)	NFC, eZone (mobile phone charging)	Far-field RFID

Table 1.1. *Characteristics of different WPT methods*

Inductive coupling (Figure 1.2(a)) is based on Ampere's circuital law and Faraday's law of induction. The first describes the magnetic field generated by an electrical current passing through a loop (coil) and vice versa; the second predicts the interaction between a time-varying magnetic field and an induced electromotive force. The efficiency of the transmitted power depends on the coupling coefficient, which depends on the distance between the transmitter and receiver. It is used for short-range applications, usually working with frequencies of a few megahertz. This method was the first WPT technology applied to real applications (Flack *et al.* 1971; Akin *et al.* 1998; Peng *et al.* 2010).

Resonant coupling (Figure 1.2(b)) has been widely applied to microwave band-pass filters (BPF) (Cohn 1958; García-García *et al.* 2004); however, it was not until 2007 when researchers from the Massachusetts Institute of Technology (MIT) demonstrated its use for WPT achieving up to 40% efficiency over more than 2 m (Kurs *et al.* 2007; Wei *et al.* 2014). A resonator is composed of a coil (L) and a capacitor (C), and the transmitted power is mainly magnetic. Resonators are currently used in smartphones and other mobile devices.

Unlike inductive and resonant coupling, WPT via radio waves (Figure 1.2(c)) uses the radiated electromagnetic field. Therefore, it uses antennas instead of coils. It is mainly used in far-field RFID systems (Shinohara 2014). It achieves lower power transfer but a much longer range.

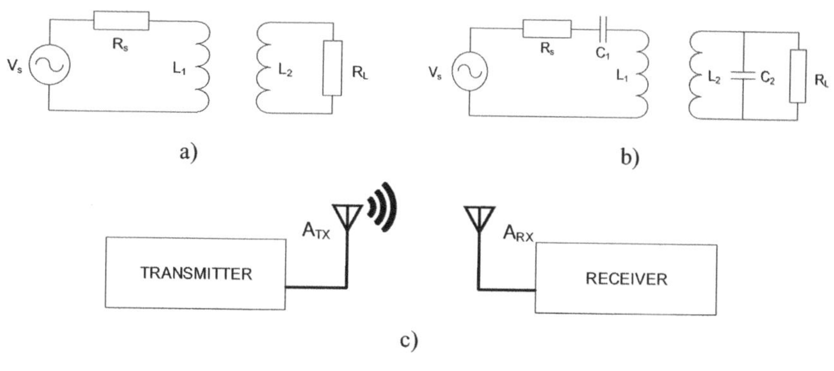

Figure 1.2. *Methods of WPT. (a) Inductive coupling; (b) resonant coupling; and (c) far field*

Resonant coupling, which is the technique used in NFC systems, relies on the principle that resonant objects exchange energy efficiently. To accomplish the power transfer, the source resonator generates an oscillating magnetic field, commonly working at the MHz range, which is received by another resonator located a few centimeters away. WPT using resonant coupling can be classified according to different parameters (Kim 2012), as shown in Table 1.2.

There are two possible ways to analyze resonant coupling. The first one involves the coupled-mode theory (CMT) (Hamam *et al.* 2007). Conversely, since resonators are based on RLC circuits (see Figure 1.2(b)), the circuit theory – which is more straightforward – can be applied as well and it is the approach used in this chapter.

Classification	Type	Description	Comments
Field used in coupling	Magnetic coupled	Affected by the surrounding permeability	Not affected by the body (medical applications)
	Electric coupled	Affected by the surrounding permittivity	
Power feeding	Direct-fed	Power source and load directly connected to the resonant structure	
	Indirect-fed	Feeding loop	Impedance matching by adjusting the spacing between the resonant structure and the loop
Resonant scheme	Self-resonant	Inductance and capacitance realized by identical structure	
	External resonant	Different structures	Typically the loop structure used for inductance and a capacitor is added

Table 1.2. *Coupled resonant WPT system classification*

1.2. Theoretical background

The underlying physical principles of electromagnetism are needed in order to understand WPT associated with NFC technology, and are briefly reviewed in this section. Inductive coupling consists of a primary coil generating a sinusoidal varying magnetic field, which induces a voltage on the terminals of a secondary coil, thus transferring power to a load. This effect is produced by the fact that a flow of current generates a magnetic field; the magnitude of this field is described as the magnetic field strength and is represented by H. The Biot–Savart law allows the H field strength and direction to be calculated from any geometry and current, for any point in space (Figure 1.3). Its integral form is given by [1.1]:

$$\vec{H} = \frac{I}{4\pi} \oint_S \frac{d\vec{s} \times \vec{r}}{|r|^3} \qquad [1.1]$$

where I is the current, $d\vec{s}$ is a vector along the path S, and r is the point in which the field is being computed.

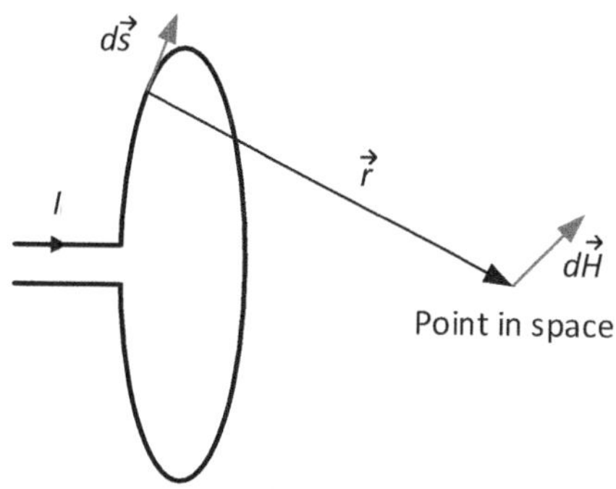

Figure 1.3. *Representation of the Biot–Savart law to measure the H field strength in any point of the space. For a color version of this figure, see www.iste.co.uk/girbau/sensors.zip*

For practical applications, the previous integral is not appropriated as closed analytical solutions only exist for special cases. The H field on the center axis of a cylindrical coil is given by (Paret 2016):

$$H = \frac{INr^2}{2\sqrt{(r^2 + x^2)^3}} \qquad [1.2]$$

where N is the number of turns, r is the radius of the loop and x is the distance from the center of the coil. Furthermore, when the loop has a rectangular shape, with side lengths a and b, the strength of the magnetic field at distance x is (Paret 2016):

$$H = \frac{INab}{4\pi\sqrt{\left(\frac{a}{2}\right)^2 + \left(\frac{b}{2}\right)^2 + x^2}} \cdot \left(\frac{1}{\left(\frac{a}{2}\right)^2 + x^2} + \frac{1}{\left(\frac{b}{2}\right)^2 + x^2} \right) \qquad [1.3]$$

The Biot–Savart law achieves accurate results for the near-field region, and it is used to obtain the homogeneity of the emitted H field and the optimum antenna radius.

The amount of magnetic field that passes through a given surface (Φ) can be measured by:

$$\Phi = BA \quad [1.4]$$

where B represents the magnetic flux density and A is the area of the given surface. The relationship between the flux density (B) and the field strength (H) is given by:

$$B = \mu_0 \mu_r H = \mu H \quad [1.5]$$

being that μ_0 is the vacuum permeability, and μ_r is the relative permeability (μ is the product of μ_0 and μ_r). When using coils of N loops, the total magnetic flux (ψ) is the sum of the flux (Φ):

$$\psi = \sum \Phi_N = N\Phi = N\mu HA \quad [1.6]$$

The relationship between the magnetic flux and the current is the inductance (L), and can be expressed as:

$$L = \frac{\psi}{I} = \frac{N\Phi}{I} = \frac{N\mu HA}{I} \quad [1.7]$$

The physical principle of NFC for communication relies on the mutual inductance between two coils (M_{21}). Mutual inductance defines the effect of the magnetic field generated by the first coil with area A_1 over a second coil, located near the first, with area A_2:

$$M_{21} = \frac{\psi_{21}(I_1)}{I_1} = \oint_{A_2}^{\infty} \frac{B_2(I_1)}{I_1} dA_2 \quad [1.8]$$

where the sub-indices 1 and 2 refer to the first and second coils, respectively, and ψ_{21} is the coupling flux which passes through both coils. Combining Ampere's law and the Biot–Savart law, thus applying the reciprocity theorem, it can be stated that:

$$M = M_{21} = M_{12} \quad [1.9]$$

And the mutual inductance is directly related to the coil inductance by:

$$M = k\sqrt{L_1 L_2} \qquad [1.10]$$

where k is the coupling coefficient.

Figure 1.4 depicts the mutual inductance (M) between two coils.

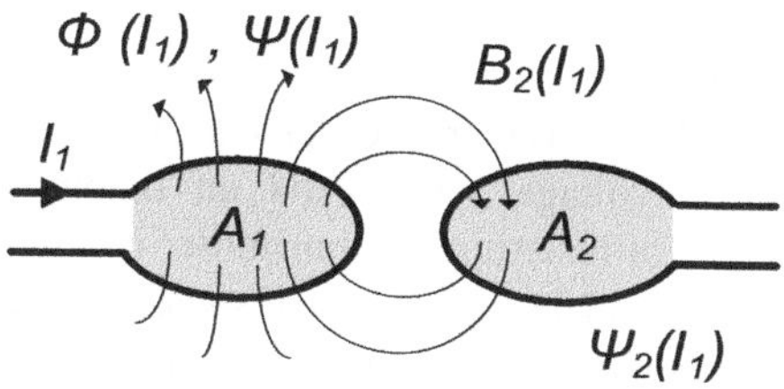

Figure 1.4. *Representation of mutual inductance (M) between coils*

This chapter analyzes WPT within the NFC scenario using a smartphone as the reader. Thus, beyond the basic background of WPT that has been explained briefly up to this point, certain specific key constraints such as the resonator design, NFC IC characterization, potential environmental effects and comparison between commercially available NFC chips with energy harvesting (EH) are addressed.

1.3. NFC systems

This section describes the reader and the tag and proposes both a system model and a procedure to understand, analyze and design WPT in NFC systems, when using a smartphone as the reader to connect to standardized NFC tags. In this context, the reader cannot, therefore, be modified and the NFC antenna depends on the smartphone used.

1.3.1. *Reader model*

Figure 1.5 shows the schema of the reader. It includes a model for the reader IC, an electromagnetic interference (EMI) filter and a matching network. The reader consists of a voltage source (V_{out}) which is connected to the EMI filter through a resistance (R_{out}). The EMI filter is a low-pass filter. Its function is reducing the second and the higher harmonics and performing an impedance transformation. This filter is composed of two resistors (R_0), two inductors (L_0) and two capacitors (C_0).

Two capacitors (C_s) are used as the matching network to match the impedance between the EMI filter and the antenna.

Two resistors (R_x) and a capacitor (C_{in}) compose the receiver circuit to adjust the received power levels, avoiding its saturation. The antenna is modeled as an inductance (L_1) in series with a resistance (R_a) and a capacitance (C_a). C_p is the parasitic capacitance of the antenna. Two resistors (R_Q) may be added between the antenna and the matching network in order to reduce the quality factor (Q) and to meet with the standard requirements.

As is also shown in Figure 1.5, the reader can be modeled using a Thevenin equivalent circuit with a voltage source and an output impedance (R_s). This simplified model is useful for inserting the transmitter in circuit simulators. A series capacitance (C_1) is added to make the circuit resonant at the operating frequency. The parasitic capacitance of the reader antenna (C_p) in the matching network has been considered and is included in the equivalent Thevenin model. An equivalent model can be used for the reader antenna, which is modeled as an inductance (L_1) and its losses (R_1, including R_Q) in series with the inductance. The reader includes a matching network, so a perfect match is assumed at the resonance frequency, $R_s = R_1$.

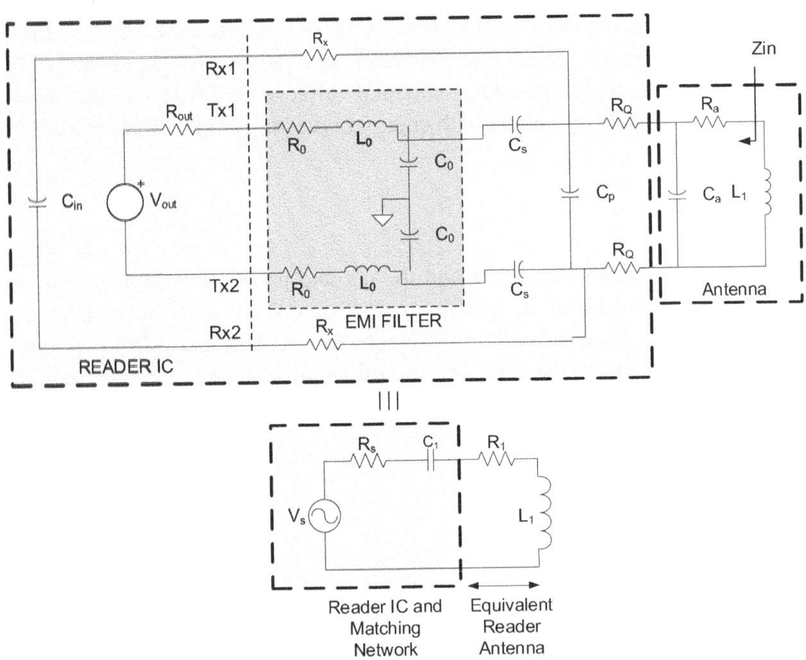

Figure 1.5. *Circuit model of the reader, including the matching network (top) and the simplified circuit model (bottom)*

It is known that in some applications the reader correctly communicates with a smart card at a large distance but not at short distances, independently of the demodulation quality of the backscattered signal (Jiang *et al.* 2005; Marechal and Paret 2008). This phenomenon is known as the loading effect, resulting in a lack of transmitted power due to the strong influence of the tag's load on the reader (Jiang *et al.* 2005; Marechal and Paret 2008).

In addition, when using a smartphone as the NFC reader, unlike other WPT systems, the transmitted power and reader antenna topology depend on the mobile model. Typical transceiver IC for NFC in commercial smartphones can achieve transmitted powers of between 20 and 23 dBm (e.g. NXP PN7120 (NXP Semiconductors 2018b) or TI TRF7970A (Texas Instruments 2014b) IC transceivers).

1.3.2. *Tag model*

Figure 1.6 shows the block diagram of an NFC tag IC. There are two main blocks connected to the antenna (Lu *et al.* 2016). The first is the WPT unit, and it has the function of harvesting the energy to power up the IC; while the second is the communication unit that demodulates the data and generates the clock for the transmission of information back to the reader. This chapter focuses on the WPT unit. This is composed of an RF limiter, a rectifier, a shunt regulator and a load modulator, which can be modeled as a shunt capacitance with the antenna.

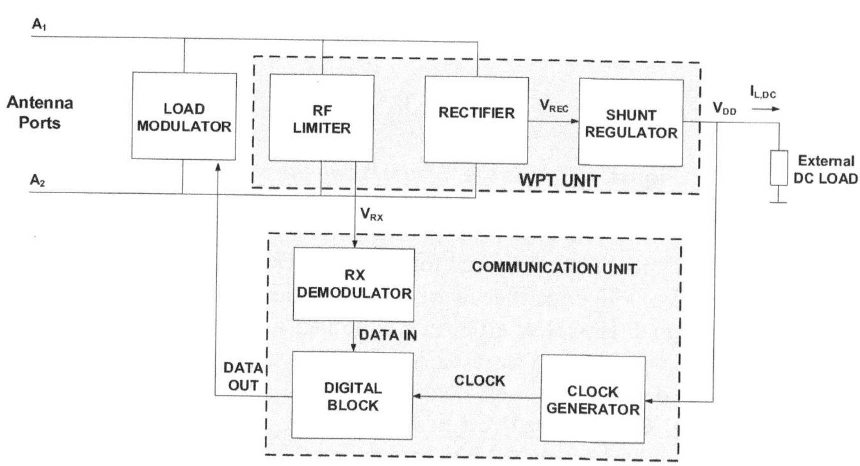

Figure 1.6. *Block diagram of the NFC tag IC*

Due to its proximity to the reader, the coupled AC signal at the antenna can be large, which could lead to the destruction of the MOS transistor. Therefore, an RF limiter is needed to prevent internal circuit damage caused by any undesired high-input signal. Then, the AC signal is translated into a DC signal by means of a full-wave rectifier. The rectified DC voltage is regulated in order to obtain a stable voltage for the operation of the communication unit.

For modeling the limiter, the circuit described in Lu, Li and Lin (2016) and shown in Figure 1.7 is considered. This limiter is based on two large-area NMOS shunt transistors that are activated when the input power is enough to produce a current through the diodes that is larger than the threshold voltage needed to activate the transistors. The limiter is also used to implement an envelope detector using the diodes and the *RC* filter (R_1 and C_1) for ASK demodulation. The output of this circuit (V_{Rx}) is connected to the input of a slicer circuit for data demodulation.

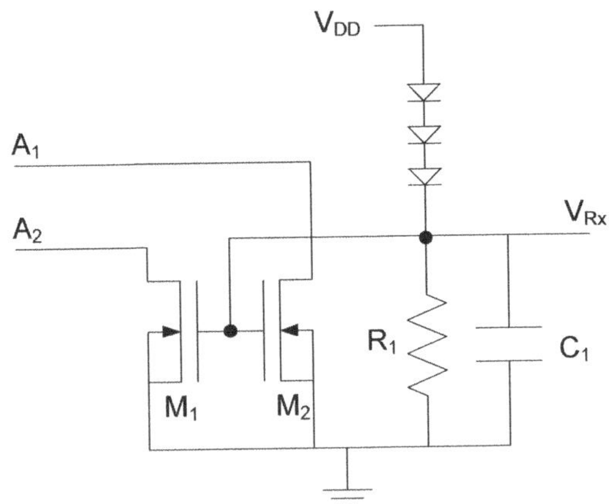

Figure 1.7. *RF limiter model inside the NFC IC*

A high-efficiency rectifier is required for the RF-to-DC conversion. This is often based on a full-wave bridge rectifier in which the diode pairs turn on during each cycle of the AC signal. However, efficiency is limited for the threshold voltage of the diodes. Schottky diodes can be used in this topology to reduce these voltage drops, but these devices are not compatible with the conventional CMOS technology. Several CMOS-based rectifiers have been proposed in the literature. The most frequently used are the NMOS bridge and the CMOS gate cross-coupled (or cross-connected) rectifier (Jamali *et al.* 2005) (Figure 1.8).

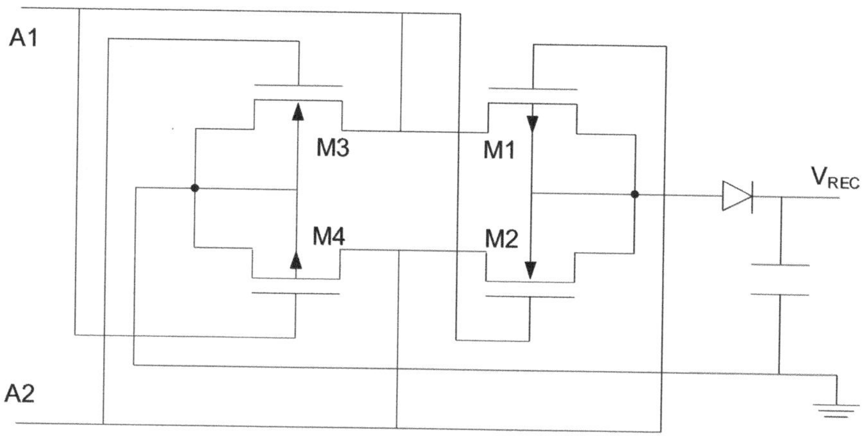

Figure 1.8. *Model of the full-wave rectifier based on the CMOS gate cross-coupled rectifier*

A shunt regulator can be used to model the regulator in the simulation (e.g. based on a zener diode). Although lateral zener diodes can be implemented in standard CMOS processes (Teichmann *et al.* 2003), the shunt regulator is often implemented with transistors. Nevertheless, all the topologies are based on the same principle which consists of reducing the shunt resistance of a bypass transistor when the input voltage is higher than the desired output voltage. Figure 1.9 shows the shunt regulator used in this analysis (Zhu 2004; Jiang *et al.* 2005; Hu *et al.* 2009). In this circuit, the shunt impedance to reduce the voltage is controlled comparing a bandgap-based reference voltage and the output voltage.

The load modulator (Figure 1.10) is formed by two NMOS transistors that act as switches, each of them connecting one capacitor from the RF input to ground (Jianguo *et al.* 2007; Alcalde *et al.* 2014). In order to transmit the information to the reader, the transistors are switched ON and OFF by the data signal connected to the gate (DATA OUT), which results in a load impedance change at the antenna as a function of the data. Since this analysis focuses on the WPT, the effect of the modulator is included in the shunt parasitic capacitance of the IC.

The proposed simplified model for the tag is shown in Figure 1.11. The tag IC is modeled as a resistance (R_{IC}) in parallel with a capacitor (C_{IC}). The tag antenna is modeled as an inductance (L_2) connected in series with a resistance (R_2). A shunt parasitic capacitance of the tag antenna (C_p) is also included. Finally, a tuning capacitance C_{tun} is added to make the circuit resonant at the operating frequency.

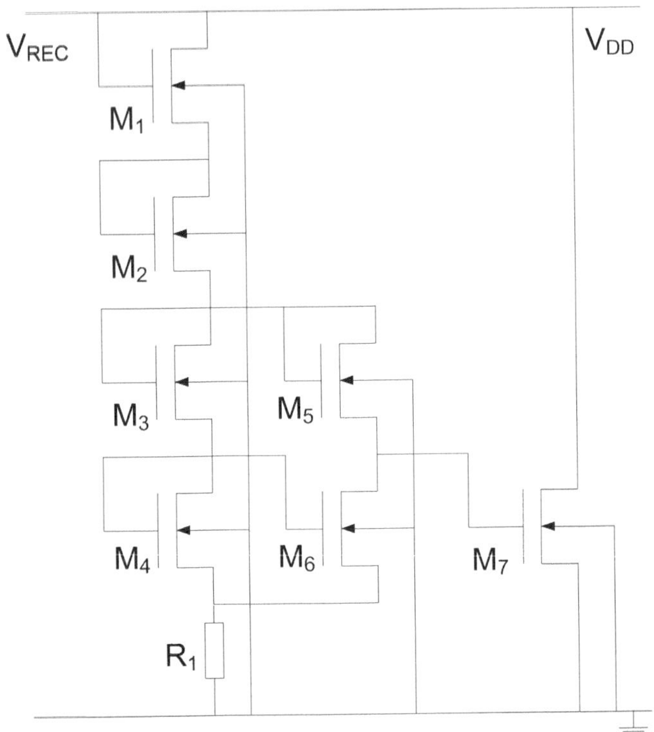

Figure 1.9. *Shunt regulator circuit model*

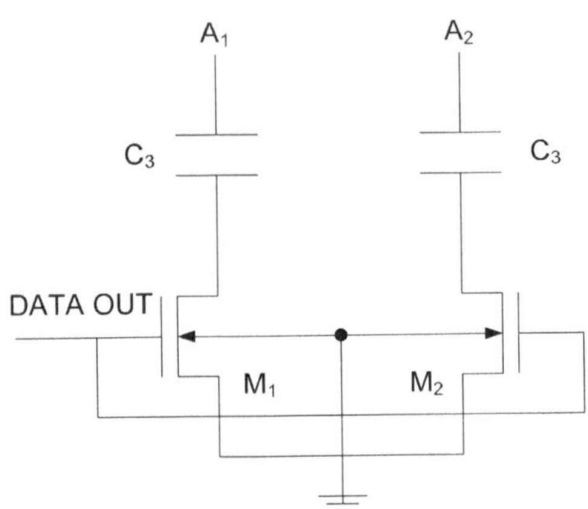

Figure 1.10. *Load modulator model formed by two NMOS transistors*

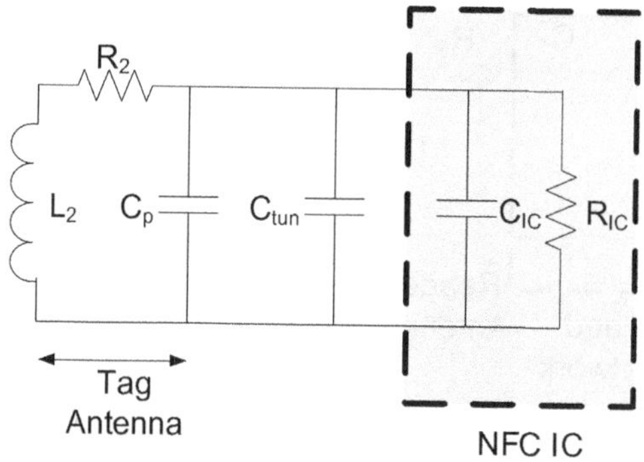

Figure 1.11. *Simplified circuit model of the tag including the tag antenna, tuning capacitor and NFC IC*

An important point to consider is that the IC impedance presents a nonlinear behavior; thus, the impedance at the operating frequency changes depending on the received power. However, it is not easy to measure this nonlinearity. The impedance of RF laboratory equipment is usually 50 Ω, and the IC of the tag is expected to have high impedance (in the order of kΩ). The second consideration at this point is the high-power level required, comparable to the real power levels generated by the reader. In Gvozdenovic *et al.* (2014), a high-power method for the impedance characterization of HF RFID chips is proposed, by using the shunt or series method, measuring the transmission coefficients (S_{21}) using a series setup. This method improves the accuracy obtained by the reflected method (measuring S_{11}) (Gebhart *et al.* 2010), since the latter is less reliable when measuring high impedance chips. As shown in Gvozdenovic *et al.* (2014), HF RFID transponders behave in a strong nonlinear way when the received power increases.

1.3.3. *NFC system model*

Figure 1.12 groups together the reader and tag models introduced in the preceding sections, showing the proposed simplified circuit model of the WPT between the reader and the tag. The reader includes the IC model followed by its matching network and antenna. The tag is modeled as an antenna, a tuning capacitor and the IC model. The energy transfer between both elements depends on the coupling coefficient (k) that will be explained later in this chapter.

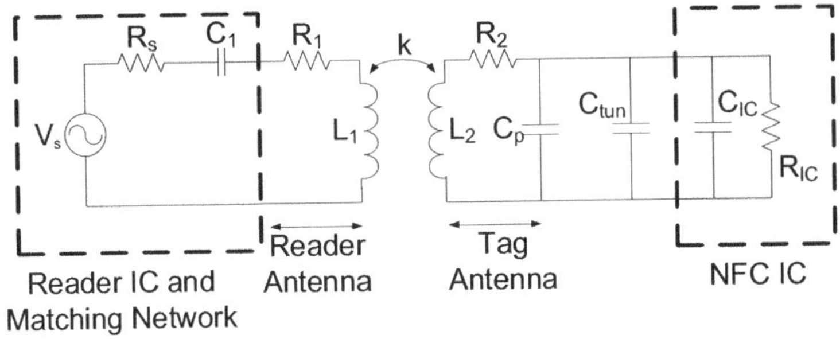

Figure 1.12. *Wireless power transfer model between the reader and the tag*

1.4. NFC constraints

Even though a device with NFC technology (reader) can transmit power wirelessly to power up a receiver (tag), its main purpose is the data transfer. Furthermore, with the incorporation of this technology in most of the smartphones manufactured today, the methods to apply it encompass a huge variety of solutions, with different antenna sizes and materials, as well as different emission power levels and electrical characteristics. Unlike other specific WPT systems which are specifically designed for the purpose of transferring power wirelessly, a number of additional restrictions must be taken into account when developing NFC devices, thus allowing an efficient power transfer and ensuring a correct data transfer, regardless of the reader used, which cannot normally be modified (especially when using smartphones), and imposes several restrictions that must be taken into account during the tag and system design.

1.4.1. *Antenna design*

As in any wireless communication system, NFC needs an antenna to establish communication between the transmitter and the receiver. In addition, when WPT is considered, not only to feed the tag for communicating but also for feeding additional electronics and sensors, it is crucial to ensure the maximum power transfer through the antennas. This section describes the main parameters to consider for their design.

1.4.1.1. *Resonance frequency (f_r), operation frequency (f_o) and antenna tuning*

The NFC nominal operation frequency (13.56 MHz) is denoted as f_0. The resonance frequency (f_r) of the antenna is the frequency at which the inductive and capacitive reactances cancel each other out, thus behaving as purely resistive. Note that, in some cases, the angular frequency (ω) is also addressed, which is related to the frequency as $\omega = 2\pi f$. To optimize power and data transfer, it is essential to adjust the resonance frequency (f_r) of the antenna to the operation frequency (f_0) of the system; this process can be done by tuning the antenna. However, it may be necessary to tune it to a slightly higher frequency when considering the change in frequency caused by environmental conditions, such as the tag material or the presence of metallic objects, as for instance a metallic reader enclosure, when working in very short ranges. The presence of a metal near the coil produces eddy currents in the opposite direction of the generated field, absorbing part of the power. This metal introduces a capacitance on the antenna impedance and reduces its inductance, which makes the energy to partially reflect at the antenna terminals since it is not matched to the input impedance.

If we take into consideration the circuit model of the loop antenna shown in Figure 1.13, the reduction of the tag antenna inductance (L_a) and changes in the parasitic capacitance (C_p, which includes parasitic antenna capacitance and parasitic capacitance due to interconnections), the resonance frequency of the antenna f_r can be written as:

$$f_r \approx \frac{1}{2\pi\sqrt{L_a\left(C_{IC} + C_p + C_{tun}\right)}} \qquad [1.11]$$

where C_{IC} is the internal IC capacitance and C_{tun} is the capacitor used to adjust the resonance frequency (f_r) to the operation frequency f_0 (13.56 MHz). The resonance frequency of a loop antenna can be measured by connecting another loop antenna to a vector network analyzer (VNA), approaching it to the loop antenna to be tested and obtaining the S_{11} parameter. The antenna connected to the VNA and the antenna to be measured should not be approached too much in order to avoid loading effects, it should be placed just close enough to be slightly coupled and to detect the resonance from the S_{11} measurement.

As mentioned before, the operation frequency of NFC is f_0 = 13.56 MHz, which leads to a wavelength of about 22 m ($\lambda = c/f_0$, where $c \approx 3 \cdot 10^8$ m/s). This leads to a half-wave dipole antenna of 11 m, which is not suitable for a handheld or a smartcard. The antenna used is much smaller; therefore, the radiation efficiency is small, and the typical parameters measured in antennas, such as the gain or the radiation pattern, are not relevant. An NFC antenna is basically a loop-shaped inductor, as shown in Figure 1.13, and its performance depends mostly on the loop area (A) and the number of turns (N). As a general rule, the larger the area and the more turns it has, the better its performance. R_p is the parasitic resistance, the inductance (L_a) is mainly defined by the number of turns of the loop, the resistance

(R_a) by the diameter and total length of the loops and the capacitance (C_p) by the separation between the antenna loops and the number of turns.

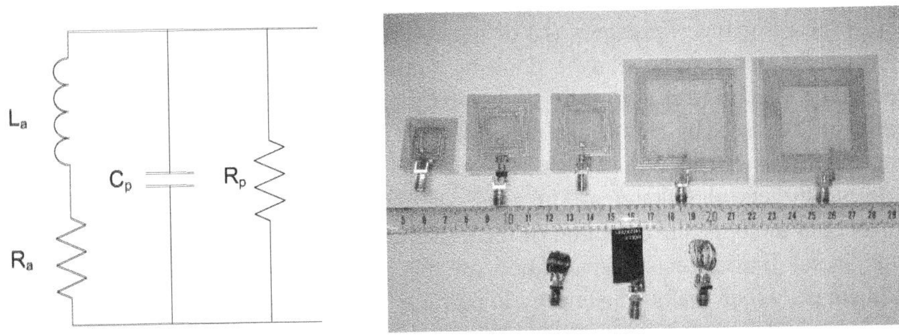

a) b)

Figure 1.13. *(a) Circuit model for a loop antenna and (b) antennas of different sizes and numbers of turns. For a color version of this figure, see www.iste.co.uk/girbau/sensors.zip*

The theoretical inductance for a rectangular wire coil can be calculated from (Jacobi and La Cost 2017):

$$L_a = N^2 \frac{\mu}{\pi}\left[-2(w+h) + 2\sqrt{h^2+w^2} - h\ln\left(\frac{h+\sqrt{h^2+w^2}}{w}\right) - w\ln\left(\frac{w+\sqrt{h^2+w^2}}{h}\right) + h\ln\left(\frac{2h}{a}\right) + w\ln\left(\frac{2w}{a}\right)\right] \quad [1.12]$$

where N is the number of turns, w and h are the average width and height of the coil, respectively, and a is the PCB width. The inductance can be measured from the Z parameters as: $L_a = Im(Z)/w_0$.

The easiest way to precisely calculate L_a and R_a is by connecting the coil antenna to a VNA and then calculating the impedance of the antenna (Z_{11}) from the measured S_{11} parameter at the operation frequency (f_0), knowing that (NXP Semiconductors 2018a):

$$Z_{11} = R_a + j2\pi f_0 L_a \quad [1.13]$$

The values of resistance and inductance are usually given by the VNA; however, the capacitance is not measured but can be estimated using the equation (NXP Semiconductors 2019):

$$C_p = \frac{1}{(2\pi f_r)^2 L_a} \qquad [1.14]$$

A series equivalent total resistance of the antenna (R_T) at the operation frequency (f_0), which combines the antenna and parasitic resistances (R_a and R_p, respectively), can be calculated by first measuring the equivalent resistance at the resonance frequency ($R_{p(fr)}$) with the VNA, and using its value to measure the parallel equivalent resistance at f_0 ($R_{p(fo)}$):

$$R_{p(f_0)} = \frac{R_{p(f_r)}1}{\sqrt{\frac{f_0}{f_r}}} \qquad [1.15]$$

and then calculate R_T using:

$$R_T = R_a + \frac{(2\pi f_0 L_a)}{R_{p(f_0)}} \qquad [1.16]$$

The parasitic resistance (R_p) can be calculated using the following expression (NXP Semiconductors 2019):

$$R_p = \frac{(2\pi f_0 L_a)^2}{R_a + 2R_Q} \qquad [1.17]$$

where R_Q is the damping resistor used to adjust the quality factor (defined and explained in equation [1.28]).

1.4.1.2. *Antenna factor (AF)*

The magnetic antenna factor (AF) at a determined frequency (f) is defined as the ratio between the incident magnetic field (H) and the output voltage at the load (V_o) (Ishii and Komiyama 2004):

$$AF(f) = \frac{H(f)}{V_o(f)} \qquad [1.18]$$

The AF of the coil antenna is used to measure the average magnetic field in the tag. In order to obtain the AF of the antenna, the circuit of Figure 1.14 can be used.

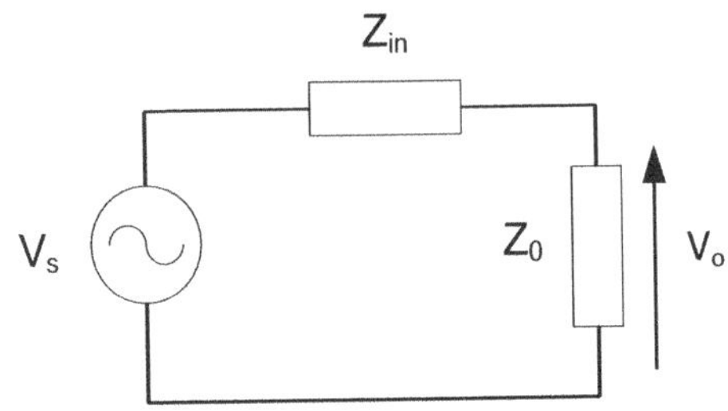

Figure 1.14. *Circuit considered to obtain the* AF

Taking into account Faraday's law: $V_S = -j\omega\pi a^2\mu_0 H$, where a is the radius of the antenna, and knowing that $V_0 = (Z_0/(Z_{in} + Z_0))V_S$, Z_{in} being the antenna impedance and Z_0 being the reference impedance (50Ω), the AF can be obtained from the measured antenna impedance as (Ishii and Komiyama 2004):

$$AF = \frac{Z_0 + Z_{in}}{j2\pi f\mu_0 Z_0 A \cdot N} \qquad [1.19]$$

where A is the loop area and N is the number of turns of the loop.

1.4.2. *Coupling coefficient* (k)

The coupling coefficient (k) of a pair of coils (C_1, C_2) is a measure of the magnetic field transferred between them and therefore it is a fundamental parameter in NFC. It depends on the shape of the antennas, the distance and alignment between them, and the materials. The coupling coefficient can be calculated, as shown in expression [1.20], from the Z parameters either obtained from electromagnetic simulation or from the S parameter measurement performed with a VNA (Zargham and Gulak 2012)

$$k = \frac{M}{\sqrt{L_1 L_2}} = \frac{\sqrt{Im(Z_{12}) \cdot Im(Z_{21})}}{\sqrt{Im(Z_{11}) \cdot Im(Z_{22})}} \qquad [1.20]$$

It is known that the maximum coupling coefficient between two coils is obtained when the radius of the coils is identical (Zierhofer and Hochmair 1996; Lazaro *et al.* 2018). However, NFC antennas (especially in the reader side) are embedded in a large variety of devices, and each of them uses a different antenna configuration. This factor must be considered when designing NFC tags.

1.4.3. *Quality factor (Q)*

The quality factor (Q) of a resonator determines its bandwidth. It depends on the relationship between the imaginary and real parts of its impedance Z. It can be expressed as:

$$Q = \frac{Im(Z)}{Re(Z)} \qquad [1.21]$$

The relationship between the quality factor at the resonance frequency (f_r) and the bandwidth (BW) can be described as:

$$BW = \frac{f_r}{Q} \qquad [1.22]$$

A high value of Q at the transmitter (reader) is translated into a narrow band, as can be seen in Figure 1.15, which benefits the power transfer. However, to ensure that the sidebands, produced by the backscattering modulation containing the information, are correctly received by the reader, the standards determine a maximum Q factor.

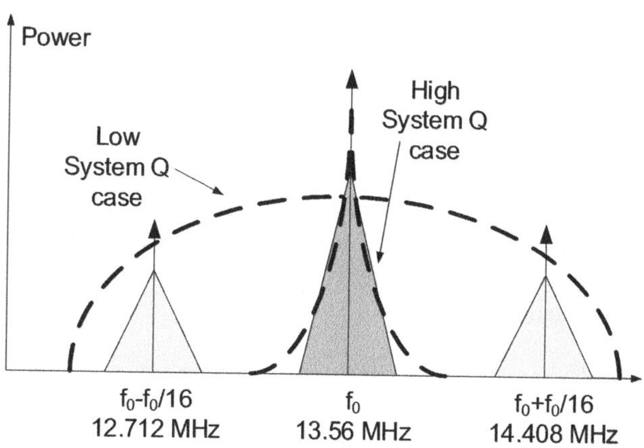

Figure 1.15. *Frequency spectrum of the NFC system with the carrier frequency (f_0) and the sidebands*

The maximal Q of a transmitter depends on the subcarrier frequency and the type of coding used. In the worst-case scenario (ISO/IEC 14443A), a maximal value of Q is $f_0/(2\ x\ bit\ rate)$, which, at a bit rate of 106 kbps, leads to a maximum Q of 64. ISO/IEC 15693 allows higher values of Q, and its maximum value comes from the pause on the carrier frequency (T_p, depicted in Figure 1.16) such as $Q_{max} = f_0 \times T_p$, being $T_p = 9.44\ \mu s$; thus, $Q_{max} = 128$. It is important to mention that the quality factor of printed antennas usually has a Q higher than that required by the standards; therefore, a series resistor is added in order to reduce it (included in R_l in Figure 1.12, which models the antenna resistance and this series resistance).

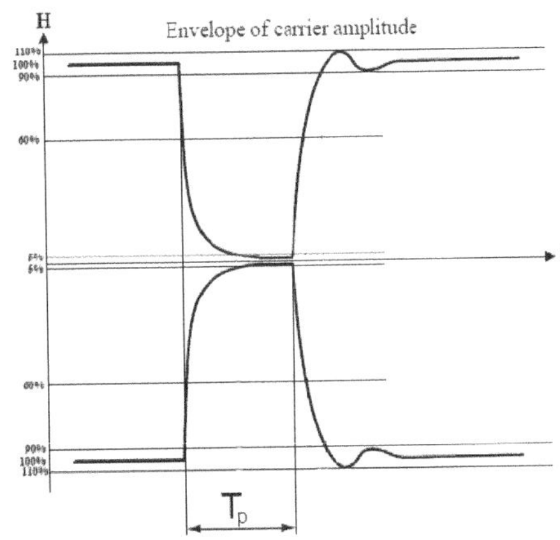

Figure 1.16. *ASK with a 100% modulation envelope amplitude. T_p is the protocol pause time, which determines the maximum* Q

Three different Q's can be defined for each side (reader and tag): the antenna quality factor (Q_i), the quality factor of the load (i.e. the NFC IC, Q_L) and the total quality factor of the whole transponder (Q_{iL}). The unloaded quality factor of the coil i (being $i = 1$ for the reader, and $i = 2$ for the tag) is obtained from the inductance and resistance at the angular resonance frequency (ω_0) with [1.21]. The total quality factor (Q_{iL}) is computed from:

$$Q_{iL} = \frac{Q_i Q_L}{Q_i + Q_L} \qquad [1.23]$$

where the external quality factor (Q_L) is:

$$Q_L = \frac{R_{IC}}{\omega_0 L_i} \qquad [1.24]$$

where $i = 1{,}2$, and corresponds to the component values in Figure 1.12.

Analogously to the case of the transmitter, the total quality factor of the tag (Q_{2L}) is limited to avoid degradation of the modulation of the subcarrier. Therefore, the maximum value of Q_{2L} is approximately $Q_{2L,max} = 8\pi \approx 25$ (Gebhart 2011). A higher quality factor would increase the time constants for the transition and decrease the load modulation. This limitation is not present in WPT systems devoted to wireless charging. The total quality factor of the tag (Q_{2L}) is given by the hyperbolic average of the quality factor of the tag antenna (Q_2) and the quality factor associated with the IC impedance (Q_L). Thus, the quality factors of the tag are defined by:

$$Q_2 = \frac{\omega_0 L_2}{R_2} \qquad [1.25]$$

$$Q_L = \frac{R_{IC}}{\omega_0 L_2} \qquad [1.26]$$

$$Q_{2L} = \frac{Q_2 Q_L}{Q_2 + Q_L} \qquad [1.27]$$

As mentioned above, the Q factor of an antenna depends on its inductance and series resistance (normally small), and, since standards determine a maximal value of Q, it is common to add two dumping resistors (R_Q) at each side of the antenna to reduce the quality factor. The value of such resistors can be calculated to reach the desired value (Q_{des}) with the following equation (NXP Semiconductors 2019):

$$R_Q = 0.5\left(\frac{\omega_0 L_a}{Q_{des}} - R_a\right) \qquad [1.28]$$

The maximum quality factor to ensure the bandwidth required for the backscattering communication (Q_{2L}) is given by:

$$R_{IC} < \omega_0 L_2 Q_{2L,max} \qquad [1.29]$$

1.4.4. *Efficiency (η)*

The efficiency (η) in WPT is defined as the ratio between the power on the load (P_L) and the emitted power (P_{in}). In the case of NFC, it depends on the coupling coefficient (k) between the coils, and the quality factor (Q) of each element, according to the following expression:

$$\eta = \frac{P_L}{P_{in}} = \frac{k_{12}^2 Q_1 Q_{2L}}{1 + k_{12}^2 Q_1 Q_{2L}} \cdot \frac{Q_{2L}}{Q_L} \qquad [1.30]$$

where k_{12} is the coupling coefficient between coils 1 and 2, and Q_1, Q_L and Q_{2L} are the quality factors of the reader antenna, the tag NFC IC and the whole tag, respectively. It is important to notice that the magnetic coupling efficiency is very sensitive to primary–secondary coil alignment and distance, since the magnetic field decays as $1/r^3$, r being the distance between coils.

1.4.5. *Average magnetic field* (H_{av})

The average magnetic field (H_{av}) is obtained from the root-mean-square voltage (V_{RMS}) calculated from the measured power (using a spectrum analyzer (SA)) at the carrier frequency from the antenna terminals. This parameter, as the name suggests, determines the magnetic field received at the antenna, and is calculated from the voltage generated on the coil terminals and the AF; thus, it depends on the electrical and geometrical parameters of the antenna. It can be obtained as:

$$H_{av}(A_{RMS}/m) = V_{RMS} \cdot |AF| \qquad [1.31]$$

1.4.6. *Environmental effects*

The magnetic and electric fields with a ground plane directly below are greatly degraded (Huang 2010). The presence of a metal close to the antenna produces eddy currents which create a magnetic field in the opposite direction, absorbing power and detuning the antenna due to a decrease in its inductance and quality factor. Considering that the most common readers used in NFC are smartphones, which usually have a metallic enclosure, or, in some cases, place the antenna over the battery surface, this situation must be taken into account. The most common way to deal with this undesired effect is to put a sheet of a high-permeability material between the antenna and the ground plane, thus creating a shield. Ferrite is the material normally used for this purpose, to help isolate the antenna. However, it does not completely eliminate the effect of the metal surface since only part of the magnetic flux will be conducted through the thin foil of ferrite, as depicted in Figure 1.17.

Another particular case is the design of tags for medical applications, where the tag is placed on the skin or is implanted. It is important to note that the inductance (L_a) is not affected by the dielectric material and is therefore unaltered when the antenna is implanted on the body. However, implantable devices must use a coating layer to isolate the electronics from the body. This is commonly done by using silicon rubber due to its high chemical inertness and durability. The capacitances introduced by the coating layer must be taken into account and this is included as C_s

in the circuit model of Figure 1.18. In addition, the series resistance (R_a) is frequency-dependent, mainly due to the skin effect. The conductivity of biological tissue is much smaller than that of the metal used for the coil strips, so the eddy current generation in the tissue is negligible.

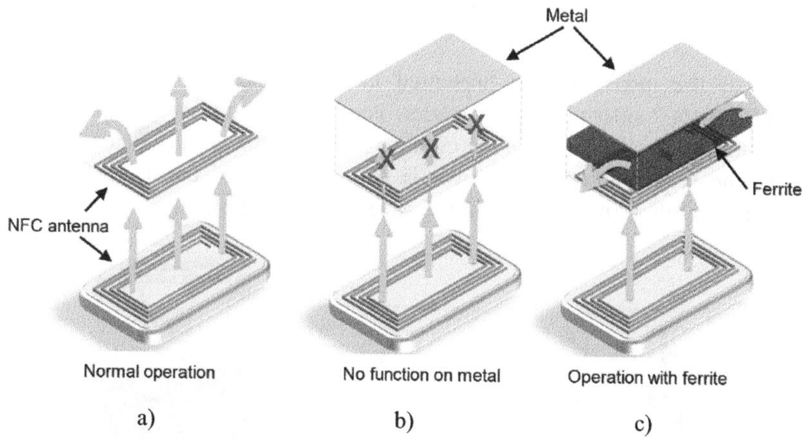

Figure 1.17. *Presence of a metal close to the antenna. (a) Normal operation; (b) field affected by a metal; and (c) use of ferrite sheet. For a color version of this figure, see www.iste.co.uk/girbau/sensors.zip*

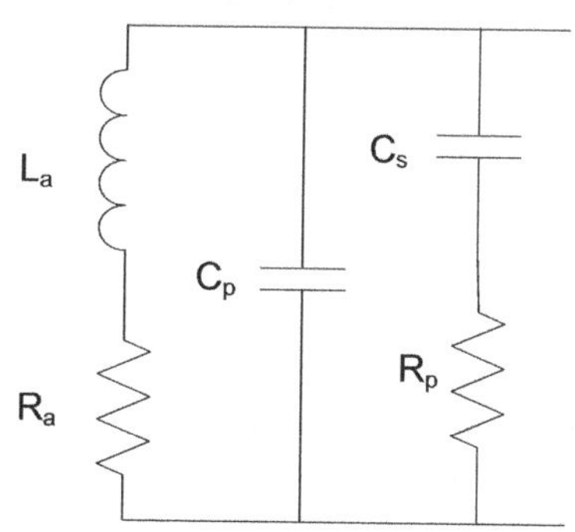

Figure 1.18. *Circuit model for an implanted loop antenna*

1.5. NFC simulation and measurement

The aim of this section is to develop the concepts and theory outlined above using simulation tools and measurement equipment. A number of practical methods are used to measure the magnetic field and the coupling coefficient. The different antenna topologies and materials are classified, as well as the different types of readers and ICs of the tag, highlighting in the latter its EH capacity, as well as the configurations used to characterize it, with the purpose of giving a general overview of the processes used for analyzing the performance and viability of the tags and applications presented in the next chapters.

Most of the measurements presented in this book have been done using a Xiaomi Mi Note 2 smartphone as a reader. The NFC antenna of this model is placed on a metallic cover that protects the circuitry of the phone and is located around the camera. The reader antenna is shown in Figure 1.19 and consists of five loops, shielded by a high-permeability material sheet (ferrite sheet).

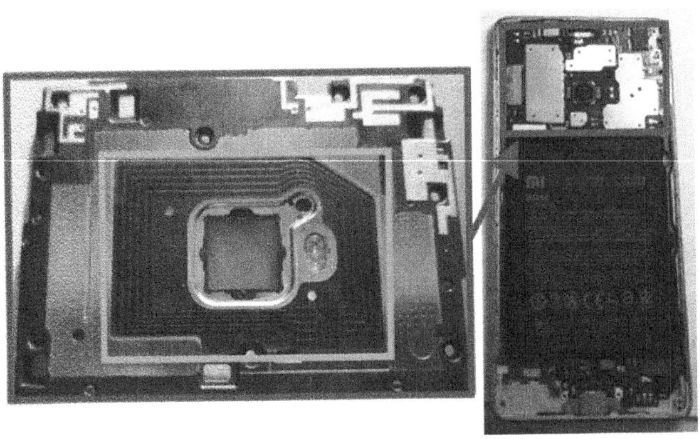

Figure 1.19. *Back of the Xiaomi Mi Note 2 (right) and the NFC antenna (green square). For a color version of this figure, see www.iste.co.uk/girbau/sensors.zip*

1.5.1. *Magnetic field* (H_{AV})

An automatized setup to measure the magnetic field received at the antenna is shown in Figure 1.20. It is composed of the reader (commercial smartphone, placed on a moving part controlled by a stepper motor, Nema 24), the antenna under test, which is connected to a vector network analyzer (VNA, Agilent E5062A) or to a spectrum analyzer (SA, Rhode & Schwarz FSP) by means of the SPDT (single-pole-double-throw) switch MSP2T-18 from Mini-Circuits, which is governed by a voltage supply (Agilent E3631A). The SA, the VNA and the power supply are connected to a computer via the general-purpose instrumentation bus (GPIB), and the stepper motor via the universal serial bus (USB). The procedure to control the instruments is programmed using a MATLAB script, first selecting the appropriate voltage on the power supply to select the switch output, choosing the VNA to measure the S_{11} parameter, and then the SA to measure the power received at the operation frequency (f_0, 13.56 MHz). Once the readings from the VNA and the SA are done, the stepper motor moves the reader 1 mm, and the process is repeated for a selected range of distances.

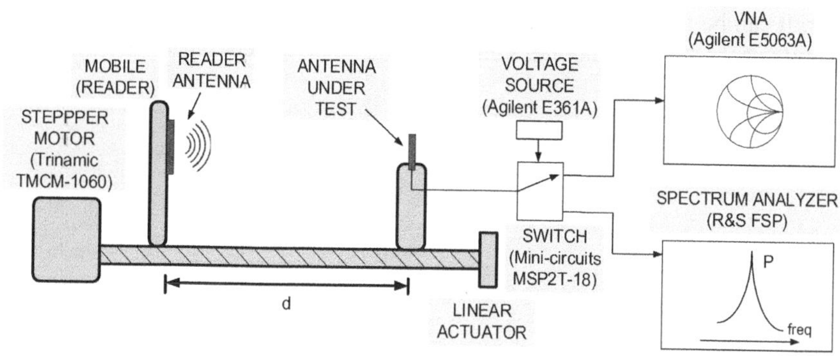

a)

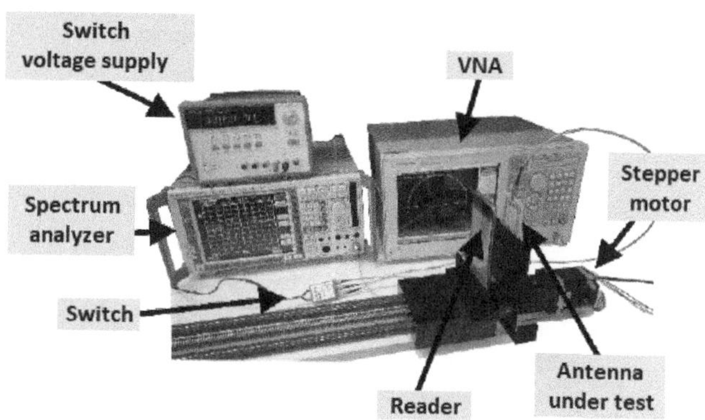

b)

Figure 1.20. *Setup to measure the magnetic field, block diagram (a) and photograph (b). The system is composed of a reader (smartphone), an antenna under test, a spectrum analyzer, a vector network analyzer, an SPDT switch and voltage supply. For a color version of this figure, see www.iste.co.uk/girbau/sensors.zip*

The performance is analyzed by measuring the reflection coefficient (S_{11}) of the antenna at each distance (d), which is used to calculate the AF. The AF is multiplied by the power received (V_{RMS}) (Figure 1.21(a)) to measure the received magnetic field (H_{av}) (Figure 1.21(b)). From the S parameters, it is possible to calculate the Z parameters which are used to calculate the inductance (L_2) (Figure 1.21(c)) and quality factor (Q_2) (Figure 1.21(d)) for different distances. The plots of these parameters shown in Figure 1.21 are obtained for a six-loop coil antenna, with an area of 50×50 mm, manufactured on FR4, and using a Xiaomi Mi Note 2 as a reader. It can be observed that the inductance and quality factor decrease due to the proximity of the metallic mobile. This effect is studied in detail in section 1.5.3.

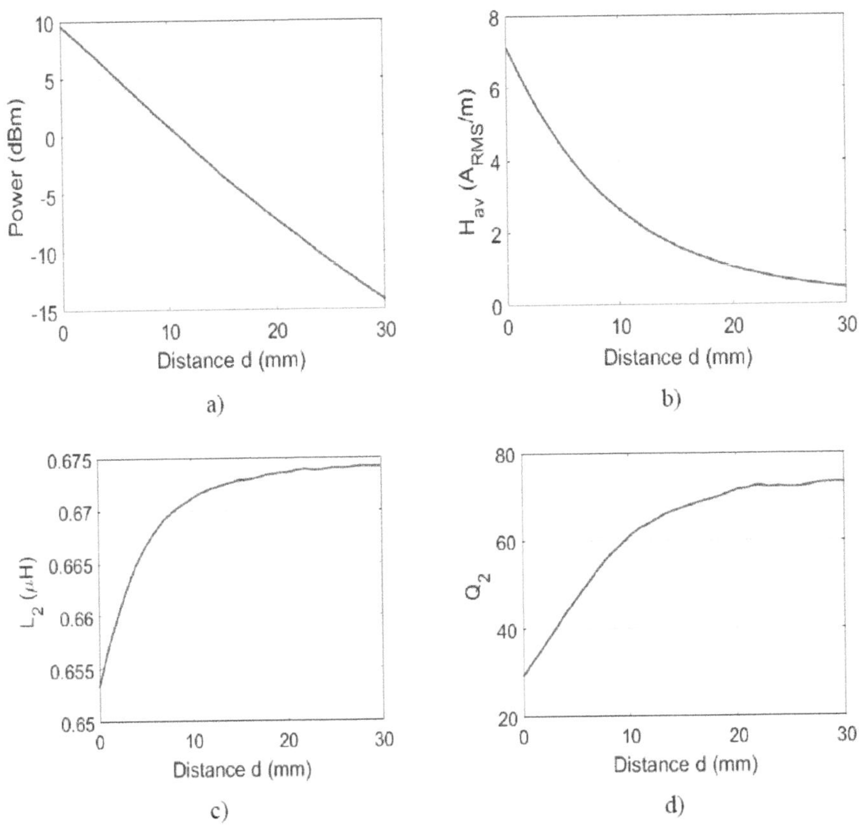

Figure 1.21. *Magnetic field measurements of a coil antenna as a function of distance using a Xiaomi Mi Note 2 as a reader. (a) Power received; (b) magnetic field received (Hav); (c) inductance (L_2); and (d) quality factor (Q_2). For a color version of this figure, see www.iste.co.uk/girbau/sensors.zip*

1.5.2. *Coupling coefficient (k)*

In order to evaluate the behavior of the coupling coefficient (k) between antennas, several coil configurations have been designed and tested in order to show a wide range of cases. Measurements have been performed with three different coils. Each coil has been manufactured on two different substrates: FR4 and Ultralam 3000 (see Figure 1.22).

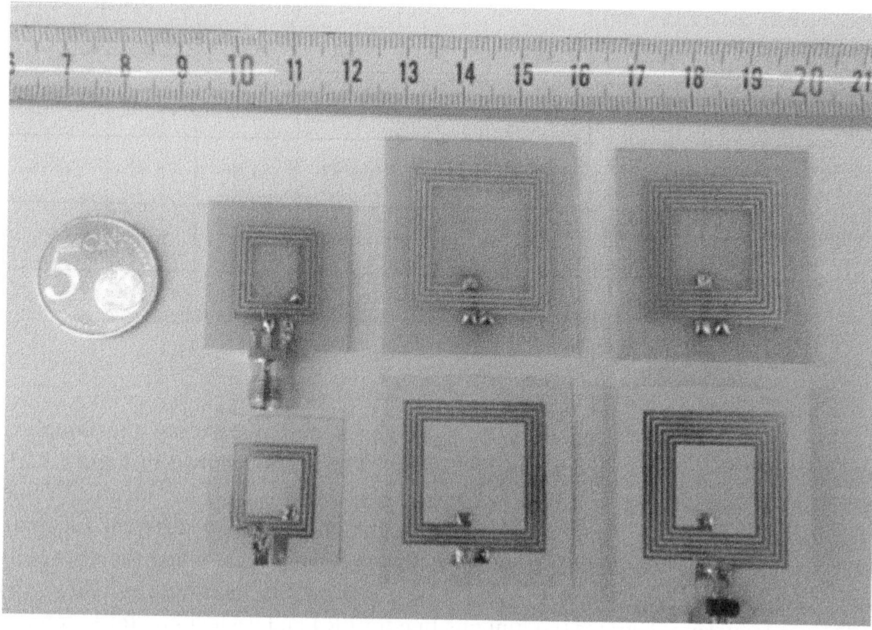

Figure 1.22. *Coils that have been designed and tested for characterizing the coupling coefficient. Coil 1 (left), coil 2 (center) and coil 3 (right). The three at the top are printed on the FR4 substrate and the three at the bottom in Ultralam. For a color version of this figure, see www.iste.co.uk/girbau/sensors.zip*

The main parameters of each coil antenna are shown in Table 1.3. Physical design parameters and measured electrical parameters are provided. Table 1.3 can be used to compare the influence of the substrate and to compare the behavior of identical coils when manufactured on the FR4 substrate versus a flexible substrate (i.e. Ultralam, UL).

	Coil 1		Coil 2		Coil 3	
	FR4	UL	FR4	UL	FR4	UL
Size (mm)	15×15		25×25		25×25	
Turns (N)	6		4		6	
Printed sides	2		1		1	
Trace width (mm)	0.7		0.7		0.7	
Spacing between traces (mm)	1		1		1	
Inductance (L_2, nH)	683	757	724	697	1182	1154
Resonance frequency (f_r, MHz)	106.33	66.57	115.83	130.29	96.69	104.31
Quality factor at 13.56 MHz (Q_2)	58.88	41.16	66.57	51.86	75.92	56.55
Antenna factor at 13.56 MHz (AF)	0.96	1.03	5.98	5.86	5.63	5.53

Table 1.3. *Coil parameters. Three different configurations, in two different substrates, FR4 and Ultralam (UL)*

Coil 1 in FR4 has been used as a reference coil, measuring the coupling coefficient (k) between this coil and the others. The results, shown in Figure 1.23, demonstrate that the best coupling is found when the two coils are identical (blue continuous line). When using the same configuration but with a different substrate (blue dashed line), a degradation on k is observed. However, when the two coils have different parameters (coils 2 and 3, red and orange lines, respectively), the difference between substrates (continuous line for FR4 and dashed line for Ultralam) is almost imperceptible.

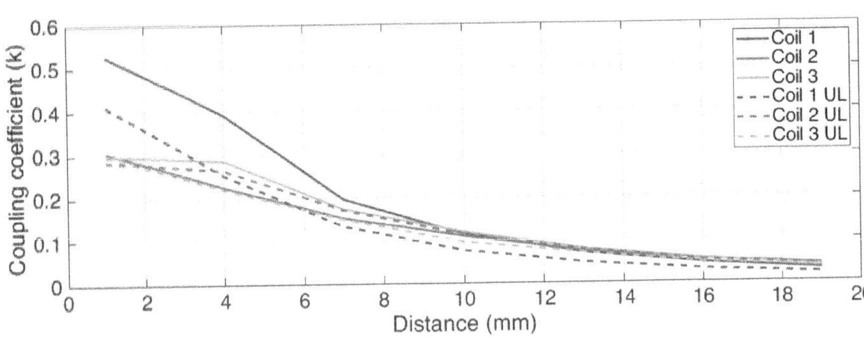

Figure 1.23. *Coupling coefficient (k) between coil 1 and six different coils. For a color version of this figure, see www.iste.co.uk/girbau/sensors.zip*

1.5.3. *Presence of metal near the antenna*

As explained in section 1.4.6, the presence of metal near the antenna produces undesirable effects. Figure 1.24 compares the measured inductance (L_2) and quality factor (Q_2) at 13.56 MHz as a function of the distance to a metallic plane that simulates the effect of metallic parts of the mobile phone. The coil antenna used for this test is a 50×50 mm with six loops, with a 0.7 mm trace width and space between traces of 1 mm. It is important to highlight that the inductance decreases when the loop antenna is close to metal, due to the induced image currents with opposite directions that reduce the magnetic flux. For distances longer than 15 mm, the inductance is nearly flat with an average value of around 1.8 μH. As the resonance frequency of the tag can be tuned adding a capacitor in parallel with the antenna, depending on the chosen value, the tag can be designed to improve its performance in ultra-short, short or long ranges. Usually, the selected distance for tuning is greater than 15 mm, because the coupling coefficient is lower than in the other cases and, therefore, the power received at the tag is expected to be higher as it is close to the smartphone.

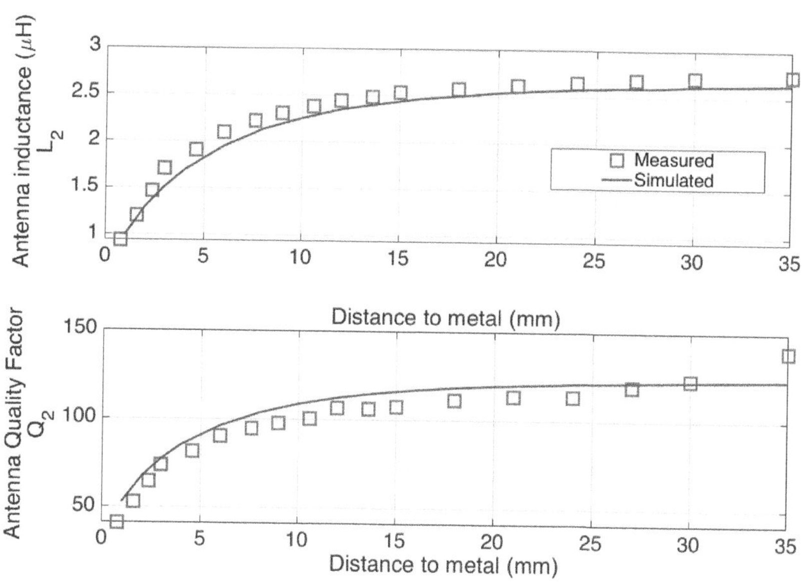

Figure 1.24. *Measured (□) and simulated (solid line) inductance (L_2) (a) and quality factor (Q_2) (b) of the loop antenna as a function of the distance to a metal plate. For a color version of this figure, see www.iste.co.uk/girbau/sensors.zip*

This effect at very short distances is compensated by the higher power received. The detuning of the antenna caused by the presence of metal is shown in Figure 1.25.

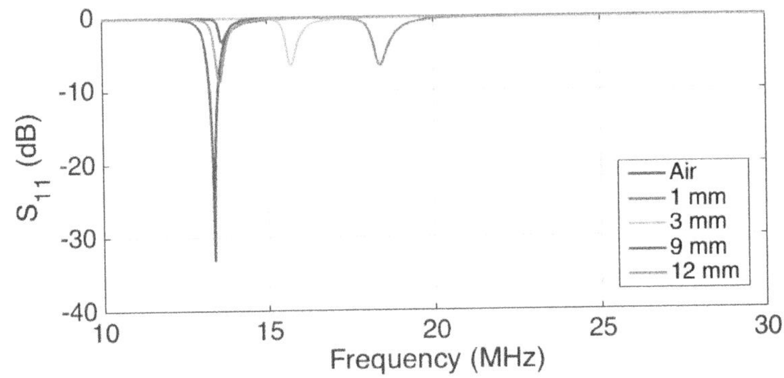

Figure 1.25. *Measured S_{11} of the antenna for different distances between the tag and the smartphone. For a color version of this figure, see www.iste.co.uk/girbau/sensors.zip*

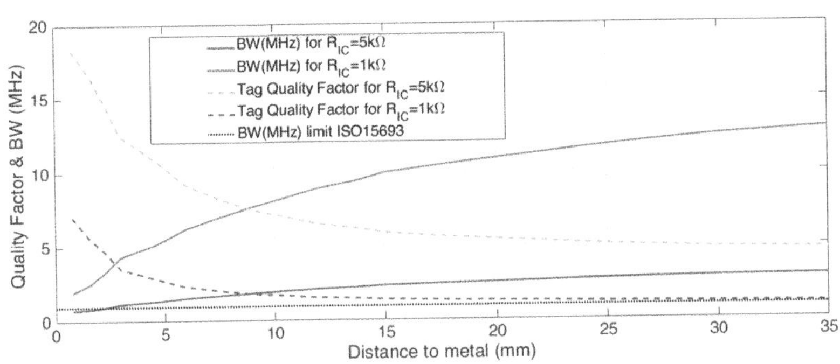

Figure 1.26. *Tag quality factor and bandwidth (BW) as a function of distance to metal depending on the IC resistance R_{IC}. The limit BW for ISO 15693 is also shown. For a color version of this figure, see www.iste.co.uk/girbau/sensors.zip*

In addition, when the tag is close to the smartphone, the bandwidth of the system increases because the tag quality factor decreases. This is caused both by the antenna quality factor reduction and by the decrease in the load resistance of the IC (due to higher power received at the tag). It is important to note that, due to power limitation of the standard VNA (often < 20 dBm), the field generated with the lab equipment is smaller than in a real reader, as is the measured field. Therefore, the IC impedance (especially the equivalent resistance) is higher than in a real situation. The IC resistance decreases typically from R_{IC} = 5 kΩ to 1 kΩ under high power excitation when the tag is very close to the reader (Gvozdenovic *et al.* 2014). Therefore, the quality factor of the tag decreases, and the tag bandwidth (*BW*) increases. Figure 1.26 shows the quality factor and *BW* as a function of the tag to smartphone distance showing that the *BW* is higher than the required ISO/IEC 15,693 (968 kHz). The shape distortion is due to the metal present in the reader.

1.5.4. *Energy harvesting*

The average magnetic field (H_{av}) received at the NFC tag must be above a threshold value (H_{min}) for a correct conversion from RF to DC. H_{min} depends on the tag resonance frequency and is obtained as (Gebhart 2011):

$$H_{min} \approx \frac{\sqrt{\left[1-\left(\frac{f_{op}}{f_r}\right)^2\right]^2+\frac{1}{Q_{2L}^2}}}{2\pi f_{op}\mu_0 AN}\cdot U_{min} \qquad [1.32]$$

where U_{min} is the minimum voltage required for the tag operation, which depends on the IC design and technology. It is important to mention that H_{min} changes depend on the IC, since each IC has different U_{min} and resistance R_{IC}. H_{min} also depends on the coil size (A) and on the number of turns (N). Considering that the tag is tuned ($f_r = f_0$), and that $Q_L \ll Q_2$, then it can be considered that $Q_{2l} \approx Q_L = R_{IC}/\omega_0 L_a$. Thus, under these approximations, H_{min} can be estimated as:

$$H_{min} \approx \frac{1}{\omega_0\mu_0 A\cdot N}\cdot\frac{U_{min}}{R_{IC}}\cdot\omega_0 L_a = \frac{I_{min}}{\omega_0\mu_0}\cdot\left(\frac{L_a}{A\cdot N}\right) \qquad [1.33]$$

Hence, the minimum magnetic field to activate the EH depends on the minimum threshold current of each chip (I_{min}), and the antenna parameters (L_a, A, and N).

The average magnetic field (H_{av}) depends on the power transmitted by the reader and the coupling between the loop antennas (distance). On the other hand, H_{min} can be considered a figure of merit of the tag because it depends only on its parameters (IC and antenna). Therefore, it can be employed to evaluate the tag performance regardless of the reader used to generate the field. Figure 1.27 shows the calculated H_{min} considering Q_{2L} = 70, L_2 = 0.7 μH, A = 15×15 mm, N = 6 and U_{min} = 4.8V, for two different values of R_{IC} (450 Ω and 525 Ω) (Gvozdenovic *et al.* 2014).

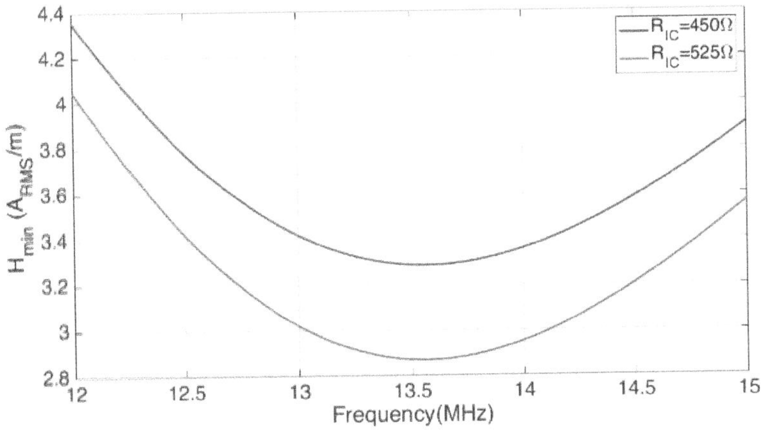

Figure 1.27. *Calculated minimum magnetic field* H_{min}(A$_{RMS}$/m) *as a function of the frequency. For a color version of this figure, see www.iste.co.uk/girbau/sensors.zip*

Figure 1.28 shows the H_{av} as a function of the distance for two smartphones. The measurements have been done with a tag on the FR4 substrate, with an antenna with six loops and an area of 50×50 mm (the same used in section 1.5.3) connected to an MR24LR NFC IC. The NFC IC is connected via an I^{2}C bus to a load consisting of an ATtiny85 microcontroller. Taking into account the current consumption of the tag, the reading range varies between 1 and 1.75 cm, depending on which smartphone is used as the reader. This distance corresponds to a measured minimum magnetic field (H_{min}) equal to 1.1 A$_{RMS}$/m (black dashed line in Figure 1.28). At the threshold distance, the measured power performed with the SA (that corresponds to the tag loaded with R_L = 50 Ω) is approximately 3 dBm. It can be seen that once the magnetic field drops below the threshold, the voltage output falls to zero (Boada *et al.* 2018b). Mobile 1 transmits higher power than mobile 2 (it is an older model) and consequently the read range is noticeably higher.

Another factor to be considered is the stability of the rectified voltage generated from the harvested energy. To ensure correct communication within the tag, Figure 1.29(a) shows the energy harvested when illuminated with a smartphone, and Figure 1.29(b) shows the data communication between the microcontroller, the tag and the sensors using the I^{2}C protocol. The result shows the viability of using digital communication while being powered up by the reader.

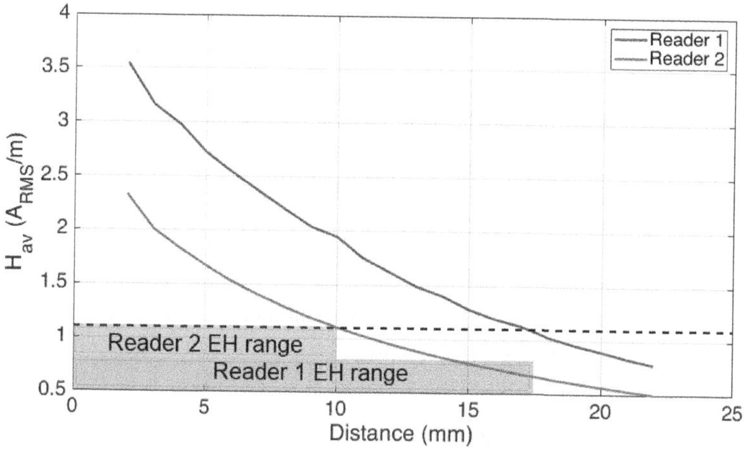

Figure 1.28. *Measured average magnetic field as a function of the tag-to-reader distance for two mobile models. For a color version of this figure, see www.iste.co.uk/girbau/sensors.zip*

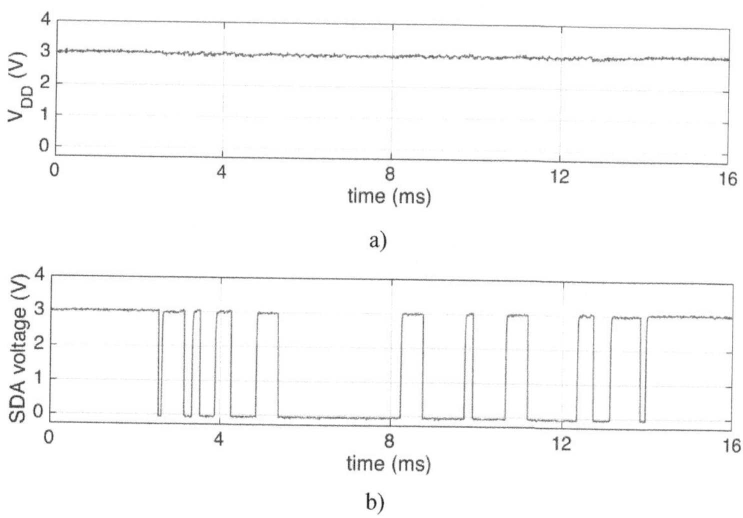

Figure 1.29. *Measurement of the energy-harvesting output (V_{DD}) (a) and the data of the SDA line from the I^2C bus (b) when the tag is powered by the NFC reader. For a color version of this figure, see www.iste.co.uk/girbau/sensors.zip*

1.5.5. *Backscattering measurement*

When a tag is powered up by receiving a magnetic field generated by the reader (H_{av}), it modulates the amplitude of the carrier frequency (f_o) with the information contained in the NFC IC memory. This modulation produces two sidebands that contain the information. Even though the correct reception of the information can be checked on the reader after reading the information, it is interesting to analyze the power of this sideband as they are the signals to be demodulated at the reader to retrieve the information. In order to measure it without affecting communication, a test coil is placed behind the tag at a distance of 10 mm, connected to an SA (Figure 1.30). Since the tag emits in both perpendicular directions symmetrically (forwards and backwards), the test antenna receives a signal very similar to the signal received at the reader.

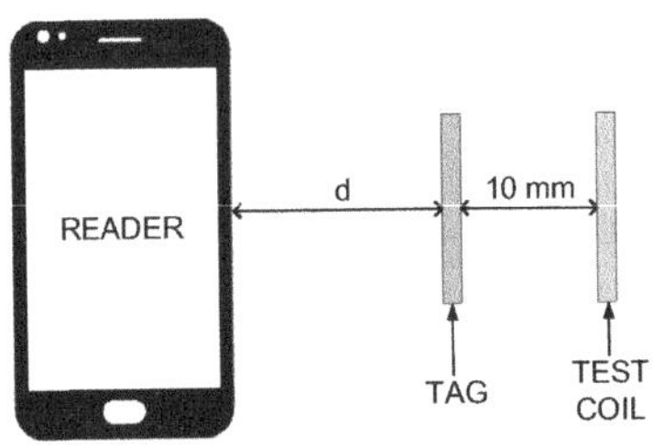

Figure 1.30. *Setup for backscattering measurement. For a color version of this figure, see www.iste.co.uk/girbau/sensors.zip*

Figure 1.31 shows the measured spectrum when the distance between the reader and tag (d) is 10 mm. The center (carrier) frequency is located at 13.56 MHz and the sidebands have a deviation of ± 424 kHz referred to the carrier (tags based on the ISO/IEC 15693 standard, where $f_{sidebands} = f_o \pm f_o/32$). The setup used for this test integrates a Xioami Mi Note 2 as the reader, and a tag based on an M24LR04 NFC IC, with a 15×15 mm coil antenna (coil 1 in Table 1.3).

The measured power of the carrier, of the lower and the upper sidebands as a function of distance is shown in Figure 1.31. It can be observed that the sidebands have different levels of power, which is because the resonance frequency (f_r) is not exactly centered at 13.56 MHz, but at a slightly lower frequency. Furthermore, at very short distances, the sidebands are hugely degraded due to the reduction of the chip resistance under high power excitation. The sideband power for the presented setup is under the noise floor for distances longer than 25 mm.

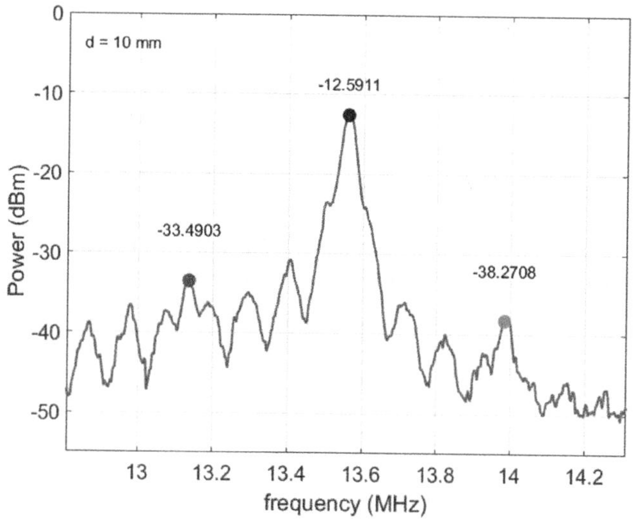

Figure 1.31. *Spectrum of the signal generated by the tag, at a 10 mm separation between the reader and tag. For a color version of this figure, see www.iste.co.uk/girbau/sensors.zip*

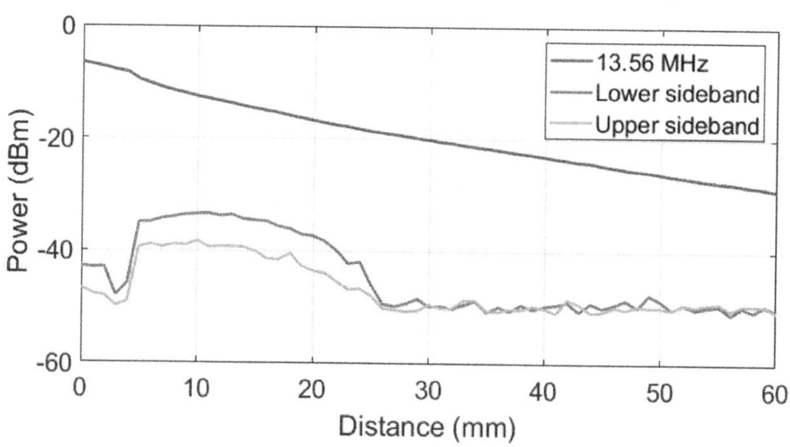

Figure 1.32. *Measured power of the carrier frequency (blue) and the sidebands (red and orange) as a function of the distance. For a color version of this figure, see www.iste.co.uk/girbau/sensors.zip*

1.5.6. *NFC IC comparison*

This section presents viable options for developing applications in real-life scenarios, hence the importance of analyzing and comparing the current, commercially available NFC IC. Table 1.4 shows a comparison of some selected NFC ICs with EH capability, providing information on the maximum current available and the typical voltage supplied after AC–DC rectification.

Manufacturer	IC model	EH max. current (mA)	Typical voltage (V)	Standard (ISO/IEC)
ST Microelectronics	M24LR-E-R (ST Microelectronics 2017b)	6	3	15,693
	ST25DV-12C (ST Microelectronics 2017a)	6	3	15,693
NXP	NT3H1101 (NXP Semiconductors 2015a)	5	2	14,443A
	NT3H1201 (NXP Semiconductors 2015b)	5	2	14,443A
EM Microelectronics	NF4 (EM Microelectronics 2014)	5	3.6	14,443A
Giantec Semiconductor	GT23SC6699 (Giantec Semiconductor 2011)	NA	3.2	14,443A
Silicon Craft	SIC4310 (Silicon Craft 2014a)	10	3.3	14,443A
	SIC4340 (Silicon Craft 2014b)	10	3.3	14,443A
	SIC4341 (Silicon Craft 2014c)	10	3.3	14,443A
AMS AG	AS3953 (AMS 2013)	5	2	14,443A
	SL13 (AMS 2014)	4	3.4	15,693
Melexis	MLX90129 (MELEXIS 2012)	5	3	15,693
Texas Instruments	RF430FRL152H (Texas Instruments 2014a)	NA	3	15,693

Table 1.4. *Comparison of NFC ICs with energy harvesting (Lazaro* et al. *2018)*

In order to perform a representative NFC IC comparison and introduce the ICs used in the tags shown in the following chapters, three different models of NFC IC with EH capability have been analyzed: two chips from ST Microelectronics (ST25DV04k (ST Microelectronics 2017a) and M24LR04E (ST Microelectronics, 2017b), both based on NFC Forum type tag 5, NFC-V, ISO/IEC 15693) and the others from NXP Semiconductors (NT3H1101 (NXP Semiconductors 2015a), type tag 2, NFC-A, ISO/IEC 14443A). For the sake of simplicity, these chips will now be referred to as ST25, M24 and NT3H, respectively. In order to compare the three ICs, they have been integrated in the same tag (see Figure 1.33). The tag integrates a 50×50 mm square loop antenna printed on FR4 (thickness 0.8 mm, metallization thickness 34 μm). The antenna has six loops, with a width of 0.7 mm and gaps of 1 mm. In all cases, the NFC IC is connected via an I^2C bus to an ATtiny85 microcontroller (Atmel 2014) programmed with the same code, which consists of a counting loop with the purpose of having the same exact load.

Figure 1.33. *Same tag with three different chips. M24 (left), ST25 (center) and NT3H (right). For a color version of this figure, see www.iste.co.uk/girbau/sensors.zip*

To determine the performance of each chip, the RMS voltage at the antenna terminals (V_{ant}) and the rectified voltage from the RF signal (V_{DD}) have been measured as a function of the distance between the reader and tag, using an oscilloscope with a low-capacitance probe. Figure 1.34 shows these two parameters for each of the ICs. Measurements have been carried out with two different smartphones, a Xiaomi Mi Note 2 (Figure 1.34(a)) and a BQ Aquarius V (Figure 1.34(b)). Since each manufacturer uses different antennas and different ICs as a transceiver, the power emitted by the reader, the coupling between the reader and tag, and the received magnetic field change between them. Hence, the importance of measuring the parameters using different smartphones as an interrogator.

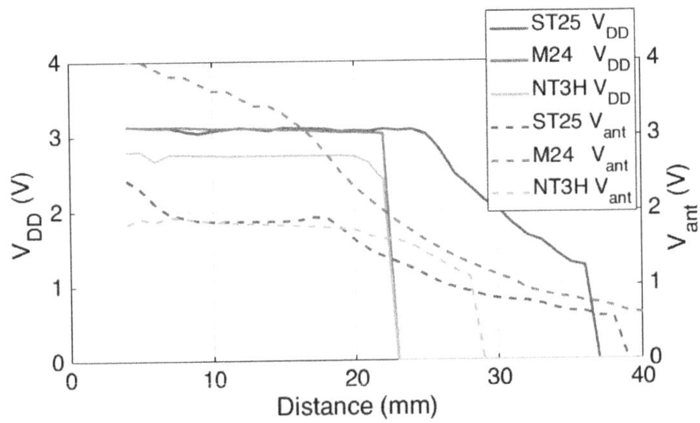

a)

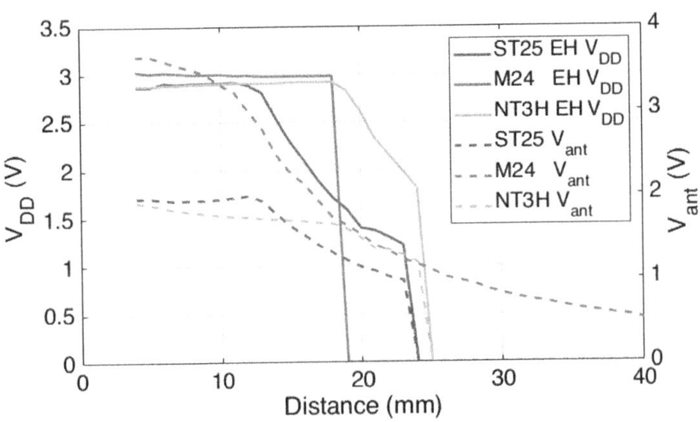

b)

Figure 1.34. *Rectified energy-harvesting output of the chip (V_{DD}, continuous line) and voltage at the antenna terminals (V_{ant}, dashed line) for different distances. Measured with two different smartphones: (a) Xiaomi mi Note 2 and (b) BQ Aquarius V. For a color version of this figure, see www.iste.co.uk/girbau/sensors.zip*

Analyzing the parameter V_{DD} (continuous line), the ST25 has a longer range; however, after a certain point (25 mm in (a), 13 mm in (b)), the output voltage degrades progressively, whereas in the case of the other two chips, an ON–OFF behavior is observed, going from about 3 V when active to 0 V abruptly. This fact might be a benefit depending on the application. For instance, some types of sensors require a certain voltage to work properly, and a smooth decrease in V_{DD} (seen in the case of the ST25 in Figure 1.34) can cause a malfunction that leads to an incorrect reading. A steep decrease of the voltage at the antenna terminals to zero (V_{ant}, dashed line) is observed with the ST25 and the NT3H. This is the result of a battery-saving operational mode of the NFC readers, which consists of sending periodic short pulses to detect the tag. If the reader cannot demodulate the answer from the backscattered sidebands of such a pulse then it stops transmitting, which means it does not detect any tag. In order to detect a nearby tag, a discovery pulse is generated every few seconds and, if detected, the pulse is held at the operating frequency, generating the voltage at the antenna terminals. Figure 1.35 depicts the signals generated by the reader to detect the tag. The graphic at the top shows the scene in the absence of the tag; approximately every 2 seconds, the reader emits a signal modulated with the different discovery requests of each supported standard. As an example, the bottom graphics of Figure 1.35 show the request modulation for NFC-B (ISO/IEC 14443B, bottom-left, using 10% ASK modulation), and for NFC-V (ISO/IEC 15693, bottom-right, using 100% ASK modulation). In the case presented in Figure 1.35, the reader has full NFC capability; therefore, the request messages for NFC-A (ISO/IEC 14443A) and NFC-F (JIS X6319) are present, as well within the discovery messages (middle-left image in Figure 1.35). This burst of requests lasts approximately 50 ms and includes polling requests of a few milliseconds each, followed by a period of unmodulated carrier to send the energy to the tag while waiting for an answer. In the midst of every attempt of polling a new tag, the reader generates short signals of the carrier frequency without modulation, shown in the middle-right image in Figure 1.35, of around 125 μs. These signals are required by the standards, and the purpose is to ensure that a listening device (i.e. a tag) will be correctly reset by a polling device (i.e. a reader), which is emitting the worst-case residual carrier level.

When the reader does not detect the presence of the tag, it keeps sending these bursts of requests, which generates a sporadic voltage at the antenna terminals (V_{ant}), but if the reader does not detect the answer to the request, it will not start sending a constant signal. Therefore, the NFC IC would not provide the rectified voltage to power up the tag. Consequently, when the voltage at the antenna terminals (Vant) maintains a value greater than zero, the reader detects the tag, even when the magnetic field intensity is below the minimum threshold required to activate the rectified voltage output of the NFC IC.

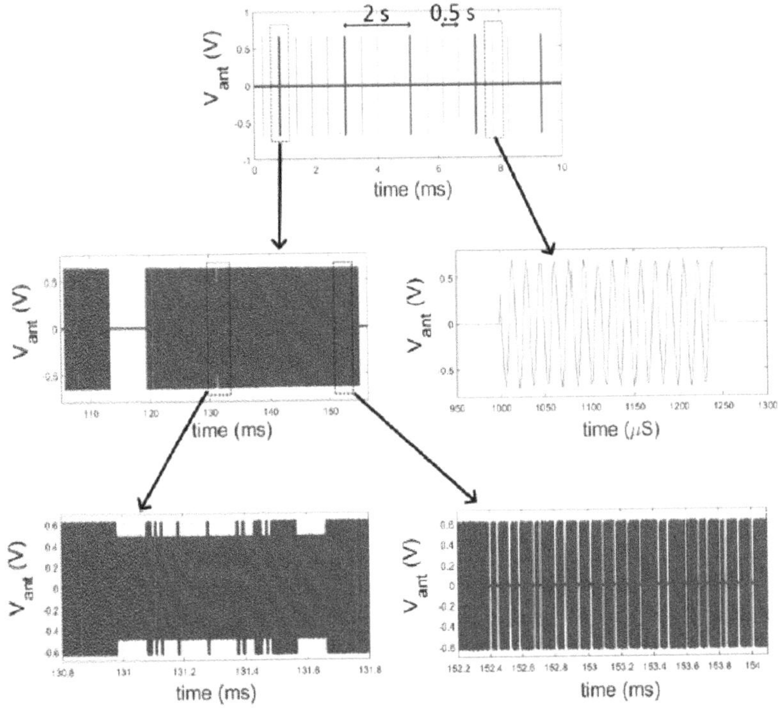

Figure 1.35. *Discovery signals sent by the reader when no tag is present. For a color version of this figure, see www.iste.co.uk/girbau/sensors.zip*

1.5.6.1. *Effects of the loads*

In order to compare the load effect at the output of the NFC IC (V_{DD}), the tags shown in Figure 1.33 have been modified, removing the MCU and connecting resistors instead. Figure 1.36 shows the measurements obtained for the M24 (Figure 1.36(a)), the ST25 (Figure 1.36(b)) and the NT3H (Figure 1.36(c)). The reader used for these measurements is a Xiaomi Mi Note 2.

Figure 1.36 shows the energy that NFC ICs can provide as a function of distance for different loads connected to their output. The smaller the load, the shorter the maximum distance and the supplied power. Considering the smallest load in this setup (510 Ω), the M24 can provide around 14 mW for distances up to 14 mm, the ST25 12 mW up to 18 mm and the NT3H 12 mW up to 20 mm. On the other hand, when connecting a 10 kΩ load, the M24 gives 1 mW up to 21 mm, the ST25 1.07 mW up to 30 mm and the NT3H 0.9 mW up to 47 mm.

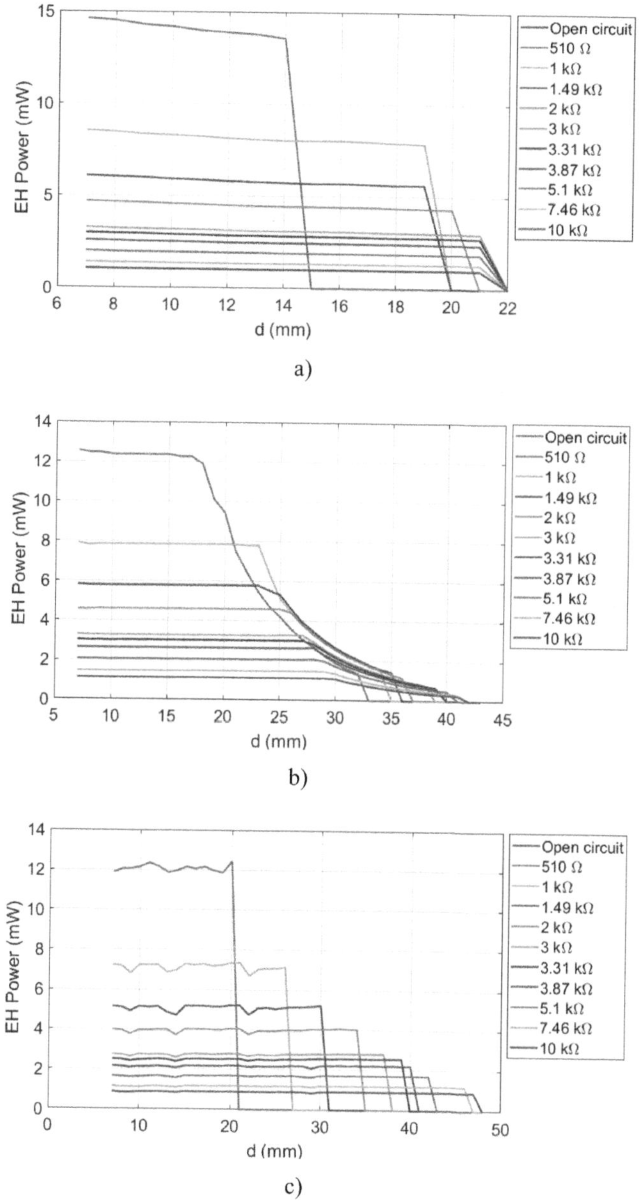

Figure 1.36. *Power harvested for different loads, with different NFC ICs. (a) M24; (b) ST25; and (c) NT3H. All the tags have a 50×50 mm six-loop antenna, and a Xiaomi Mi Note 2 is used as a reader. For a color version of this figure, see www.iste.co.uk/girbau/sensors.zip*

Figure 1.37 shows the supplied power depending on the load connected, for a distance of 14 mm between the reader and tag. It can be seen that the power decreases abruptly for loads between 0.5 kΩ and 2 kΩ, going from around 12 mW at 500 Ω to 4 mW at 2 kΩ.

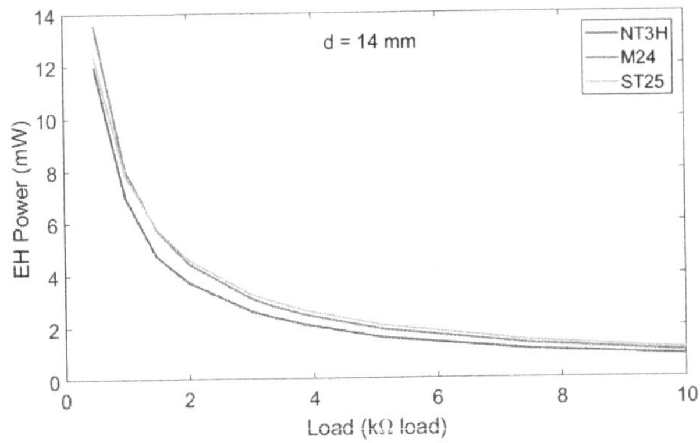

Figure 1.37. *Power at the rectified output for the M24, ST25 and NT3H ICs as a function of the load connected, at a distance of 14 mm between the reader and tag. For a color version of this figure, see www.iste.co.uk/girbau/sensors.zip*

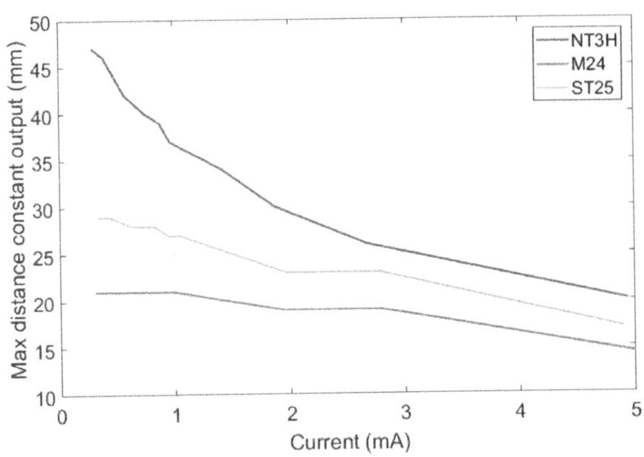

Figure 1.38. *Maximum distance at which the NFC ICs can supply a constant current for different loads for the NT3H (blue), the M24 (red) and the ST25 (yellow). For a color version of this figure, see www.iste.co.uk/girbau/sensors.zip*

From Figure 1.36, it has been deduced that the power supplied by the ICs is almost constant up to a certain point at which it drops to zero. The maximum distance at which the chips can supply certain current is shown in Figure 1.38. The NT3H IC has a longer range for identical current supply than the other two ICs. It is also the one whose current experiences greater variation with distance, since from 5 mA to 0.3 mA, it has 27 cm of difference (from 20 to 47 cm), whereas the ST25 has a 12 mm difference (from 17 to 29 mm) and the M24, 7 mm (from 14 to 21 mm).

1.5.7. *Antenna dimensions*

As mentioned previously, NFC has a wide range of possible applications. In some of those applications, the size of the tag does not matter, but in other applications, the size is essential (e.g. implantable tags). Hence, the importance of evaluating the differences between using a large or a small antenna. For this study, two antennas are compared. The first one is a 50×50 mm coil (Figure 1.39(a)), and the second is a 15×15 mm coil (Figure 1.39(b)), both with six turns (in the 15×15 mm antenna, three turns in each face of the PCB).

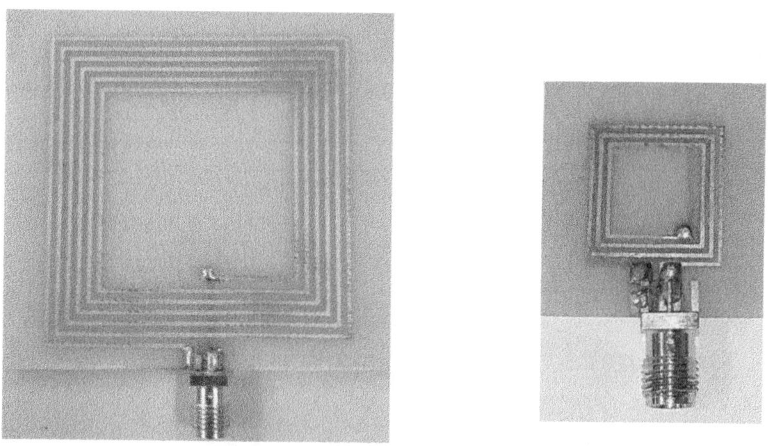

a) b)

Figure 1.39. *(a) Coil antenna with six loops printed on one side and an area of 50×50 mm and (b) coil antenna with six loops (double-sided) and an area of 15×15 mm. For a color version of this figure, see www.iste.co.uk/girbau/sensors.zip*

Figure 1.40 plots the power and the average magnetic field received in these antennas using a Xiaomi Mi Note 2 as a reader. The power is slightly higher in the case of the larger antenna because its size is closer to the size of the reader's antenna, a situation that optimizes the received power. On the other hand, the average magnetic field is higher for the smallest antenna, since, as outlined in section 1.4.5, the received magnetic field (H_{av}) depends on the AF, and the AF (defined in equation [1.19]) increases when the area of the antenna (A) decreases.

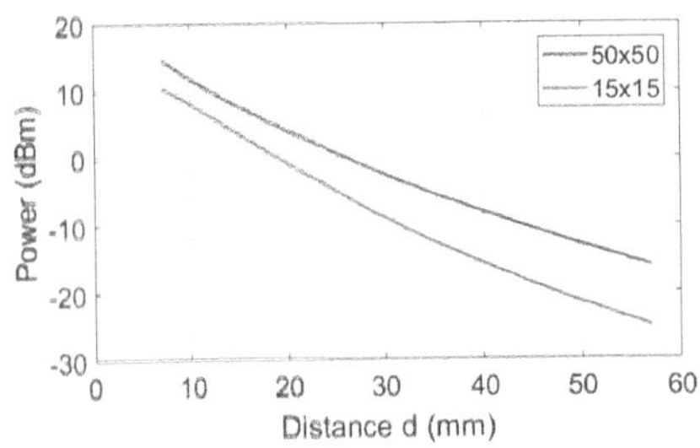

a)

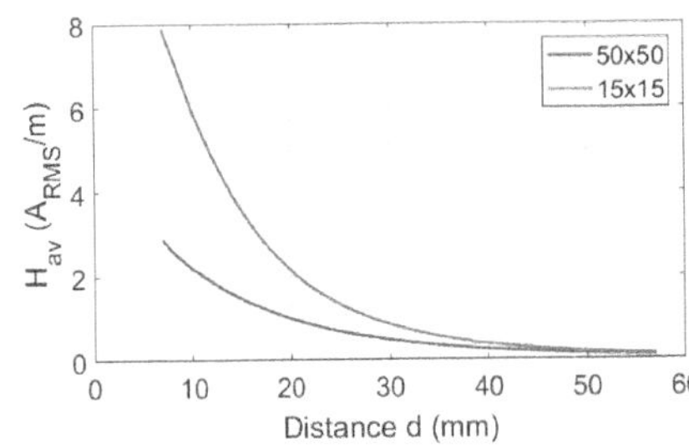

b)

Figure 1.40. *Comparison of (a) power and (b) magnetic field received by a 50×50 mm coil antenna (blue line) and a 15×15 mm coil antenna (red line). For a color version of this figure, see www.iste.co.uk/girbau/sensors.zip*

To compare the effects of the antenna size for the three NFC ICs analyzed in the previous section, the same three tags with small antenna of 15×15 mm have been made (Figure 1.41).

Figure 1.41. *Same tag with three different chips. M24 (left), NT3H (center) and ST25 (right) with a 15×15 mm antenna. For a color version of this figure, see www.iste.co.uk/girbau/sensors.zip*

Each tag consists of the antenna connected to the NFC chip, with a 3 kΩ resistor as a load which leads to 1 mA, approximately. Figure 1.42 displays the measured voltage at the load for each tag using a Xioami Mi Note 2 as a reader. The results show that the M24 (red) and the ST25 (orange) have almost the same behavior with antennas of two different sizes; however, the NT3H reduces its range by 1 cm (from 38 mm to 28 mm) when using the small antenna.

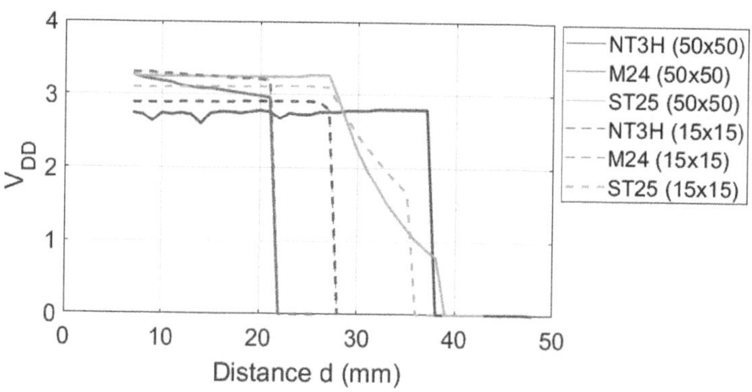

Figure 1.42. *Measured voltage (V_{DD}) for three different ICs (NT3H blue, M24 red and ST25 orange), with a 50×50 mm antenna (continuous line) and a 15×15 mm antenna (dashed line), with a 3 kΩ load. For a color version of this figure, see www.iste.co.uk/girbau/sensors.zip*

The maximum distance at which the EH is activated is obtained from the measurements shown in Figure 1.42. With those distances, and using the measurements shown in Figure 1.40(b), as well as taking into account the calculations explained in section 1.5.4, the H_{min} for each case can be estimated. The relationship between the H_{min} of the two antennas is derived from equation [1.33] and is expressed as:

$$\frac{H_{min}'}{H_{min}} = \frac{(L_2'/A'N')}{(L_2/AN)} \quad [1.34]$$

where H_{min} and H_{min}' represent the magnetic field threshold to activate the EH output of the compared antennas. Using the values of each antenna, summarized in Table 1.5, the expected ratio between the H_{min} of these two antennas is around 1.7. However, this is a theoretical value, and the real measurements give ratios of 2.1 for the M24, 1.8 for the ST25 and 3.7 for the NT3H. The difference between the theoretical and measured values is due to different factors. The misalignment between coils during the experimental measurements could lead to a mismatch between the H_{av} measured with the test coil and the H_{av} received by the tag.

In addition, even though the resonance frequency (f_r) has been carefully adjusted using the VNA, a possible factor that could induce the difference between the theoretical and the measured values is the antenna tuning, since equation [1.34] assumes that f_r is equal to the operational frequency (f_0).

	Coil antenna 50×50 mm			**Coil antenna 15×15 mm**		
	M24	**ST25**	**NT3H**	**M24**	**ST25**	**NT3H**
Area (A, mm^2)	50			15		
Turns (N)	6			6		
Inductance (L, μH)	2.74			0.683		
Minimum magnetic field ($Hmin$, A_{RMS}/m)	0.93	0.6	0.31	1.94	1.1	1.22

Table 1.5. *Comparison of* H_{min} *with three different ICs for two antennas*

2

Case Study 1: Soil Moisture Sensor

2.1. Motivation

Soil moisture measurement is a key indicator for certain applications such as irrigation water management. Irrigation regulation is essential in order to satisfy the water requirement of crops without wasting water and to properly treat the soil, controlling plant nutrients in order to save energy. Most sophisticated systems are automated using electronic climate-based controllers, whereas the simplest repeat a set of scheduled tasks at a set time. Low-cost monitoring systems are demanded for personal use at home or to monitor a particular crop, such as in a greenhouse or plant store. In these cases, irrigation is often manual or semiautomatic and the conditions are usually specific for each crop. Therefore, for appropriate plant care, soil moisture control is an important aspect. There are several soil moisture measurement techniques (Reynolds 1970) but most of them are not suitable for amateur or low-cost applications. Most typical portable low-cost meters use a sensor connected to the analog-to-digital converter (ADC) of a microcontroller, which translates a physical value to the corresponding magnitude. The result is often shown in a display or acquired by an external data logger. In all cases, these methods use a battery to supply energy to the electronic components.

Several wireless soil moisture sensors are found in the literature (Gutiérrez *et al.* 2014; Yunseop and William 2015). However, those sensors require an expensive reader (in the case of RFID systems), or batteries and expensive chips (in the case of

---

For a color version of all the figures in this chapter, see: www.iste.co.uk/girbau/sensors.zip.

Bluetooth). NFC has the advantage of its low price and the possibility of using any conventional smartphone as a reader, which can upload the sensed data to the cloud using a suitable application. One of its limitations is the short readout range (few centimeters), a characteristic which, as has been explained, defines the frame of applications and usages where NFC is very attractive and also defines the areas where this technology is restricted.

This chapter proposes, as a case study, a soil moisture sensor embedded in a passive NFC-based tag, which powers up the electronics by harvesting the energy provided by the reader. The reader is a commercial smartphone, so data obtained can be uploaded and stored in an Internet of Things (IoT) database to be used to program the irrigation schedule (also taking into account the weather forecast) and to keep track of the soil conditions. The tag also integrates ambient temperature and humidity sensors to complement the information of the environment, since these parameters may affect the irrigation schedule. The tag can be installed at several points for individual monitoring thanks to its low cost. In addition to the sensed data, an identification tag number is written into the NFC chip, which can be related to information about the plant (name, origin, brief description and specific care) which can be retrieved by the reader and displayed in a custom app.

2.2. Soil moisture measurement techniques

The Volumetric Water Content (VWC) is often used to characterize the soil moisture. It is defined as the volume of water over the total volume, as shown in the example of Figure 2.1.

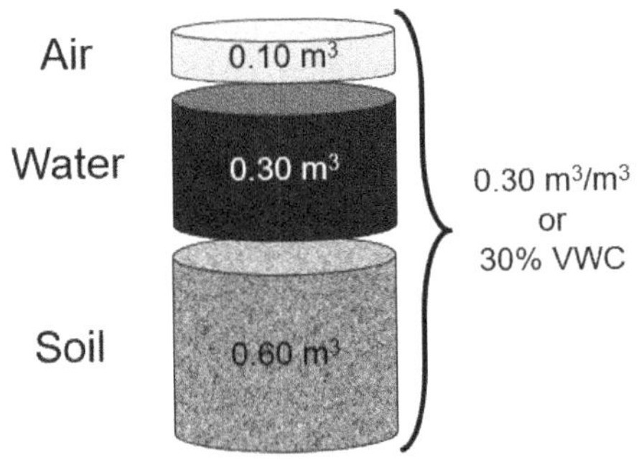

Figure 2.1. *Representation of a volume separated into constituent parts where 70% of the total volume is the soil, 10% is water and 20% is air*

Among the existing techniques of measuring the moisture content of the soil there are two types, direct and indirect. The best-known direct method is thermo-gravimetric (Reynolds 1970), which is usually used as a standard reference. It requires at least one day to get the result and it cannot be performed *in situ*. This method is therefore used for calibration of other sensors in laboratories. However, there are several indirect techniques (Robinson *et al.* 2002). The most common are listed below.

Neutron Moisture Meters (NMM) have a large action operating range and are insensitive to temperature and salinity. This is an expensive method that requires a certification for radiation (Sinclair and Williams 1979). Therefore, it can be used in research; however, it is not suitable for personal use.

Thermal sensors are based on the difference in thermal conductivity between soil and water. They measure the change in temperature of the surrounding soil after applying a heat pulse (Evett *et al.* 2003). This type of sensor is inexpensive and easy to use; however, they are highly dependent on soil temperature, present a long integration time and are delicate since the distance between the heater and the thermocouple must be kept constant.

Electrical resistance blocks or gypsum blocks (Bremner and Mulvaney 1982) are made with gypsum around a pair of stainless-steel wires. Most of the probes are based on the measurement of their electrical resistance. Portable probes are available and can be placed in the soil to give the water ratio reading. The main limitation of this technique is the calibration of the probe due to the potential corrosion of the wires with the accumulation of mineral salts.

Besides the above indirect techniques, there are other methods based on the dielectric properties of the soil in order to measure its moisture content. It is known that the electric permittivity of dry soil varies between 3 and 7 while that of pure water is around 80. Within these dielectric techniques, Time-Domain Reflectometry (TDR) is one of the most used (Jones *et al.* 2002; Cataldo *et al.* 2018). It consists of sending electromagnetic pulses along two waveguides of known length in order to measure the delay between the incident and reflected waves. The delay changes according to the dielectric constant of the soil where they sink. However, TDR systems are expensive for domestic applications or when multiple measurement points need to be controlled, requiring multiplexers to share the reading unit (Reflectometer *et al.* 2017).

The Frequency-Domain Reflectometry (FDR) (Reflectometer, Mulitplexer and Probes 2017) technique as well as the capacitive techniques (Xu *et al.* 2015) are also classified within the group of dielectric techniques because both are based on the same principle of operation. In the case of FDR, the dielectric constant of the soil is determined by applying an oscillating charge to the circuit and measuring its resonant frequency to detect variations in capacitance of the soil (Wobschall 1978; Ong *et al.* 2008). The main drawback of this technique is that the measurement of the resonant frequency requires expensive impedance meters. On the other hand, capacitive methods directly measure the capacitance, usually obtained from the capacitor charging time. Since these types of sensors are AC excited and the probe can be covered by a protective layer (Dean *et al.* 1987), they are not affected by corrosion. Nonetheless, capacitive measurements made with a microcontroller often require signal conditioning, whereas resistive sensors do not.

Despite this, a capacitive sensor is one of the best fits for the NFC scenario because printed capacitive sensors for soil measurements are compatible with standard PCB technology that is used both for the coil antenna design and for integrating NFC sensor electronics. In addition, it is not affected by corrosion issues. A critical point in using this kind of sensor is finding an accurate capacity measurement method that is compatible with the available power from the RF harvesting.

2.3. System description

2.3.1. *System overview*

Figure 2.2 depicts the system proposed in this chapter. It is composed by an NFC antenna which is connected to the NFC/RFID IC M24LR04E-R (ST Microelectronics 2017a) (M24), from ST Microelectronics, responsible for sending and receiving NFC Data Exchange Format (NDEF) messages. This device has an I^2C interface, energy harvesting with four configurable currents and is compatible with ISO 15693, which are necessary requirements for carrying out this application. When a magnetic field from a reader is received by the antenna, the NFC IC wakes up. When the energy harvesting mode is activated and the strength of the magnetic field is enough, the M24 uses the excess energy received from the RF field to power all components of the system. This chip is capable of providing currents up to 6 mA in the highest current sink mode.

The antenna is a square spiral made up of six loops printed on a PCB (FR4). Its surface is 50 mm × 50 mm and the width of the metallization is 0.6 mm. It is designed with the Keysight Momentum electromagnetic simulator. The simulated and measured inductance is 3.2 μH. In order to adjust the resonance frequency to the operating frequency at 13.56 MHz, a tuning capacitance of 15 pF is added to the internal capacitance of the IC (27.5 pF).

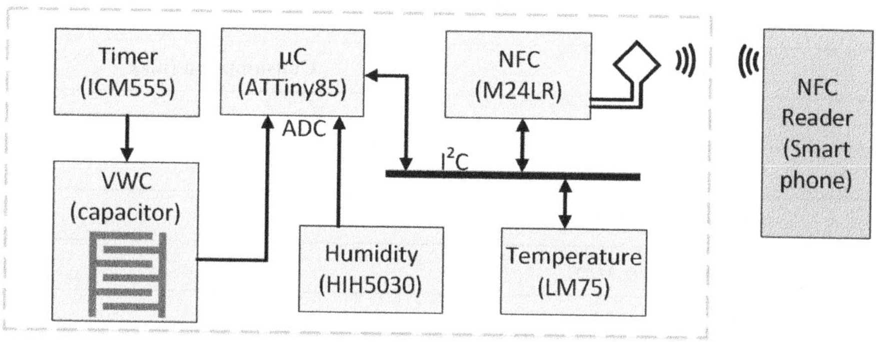

Figure 2.2. *System overview*

A low-power Atmel 8-bit AVR ATtiny85 microcontroller (Atmel 2014) has been selected to be integrated in the prototype. It can be configured at different clock speeds and can work down to 1.8 V, which reduces the clock frequency. Due to the power limitations presented by this system, it has been configured to work at 1 MHz with a current consumption around 300 µA at 3.3 V. This microcontroller can be programmed with Arduino IDE, which is very popular in open-hardware implementations. It has an I^2C interface and two analog inputs. In order to provide additional information to the user, the system not only includes a moisture sensor but also includes an ambient humidity sensor, which is connected to an analog input of the microcontroller, and a temperature sensor, which is used to adjust the moisture sensor curve. The temperature sensor is the low-cost I^2C sensor LM75A (Texas Instruments 2009), which consumes less than 280 µA in active mode and 4 µA in sleep mode (at 3 V). It is an industry-standard digital temperature sensor with an integrated sigma-delta 9-bit resolution ADC, which is enough for the proposed application. The temperature accuracy is about 1°C from -25°C to 100°C. The HIH-5030 (Honeywell 2001) is a low-voltage humidity sensor that operates down to 2.7 V, with a typical current consumption of only 200 µA. This humidity sensor presents a linear output voltage that is related to the relative humidity and an improved readout can be used if the temperature is known.

As described below, the technique used for the measurement of soil water content is based on an ICM555 timer oscillator, which is a low-voltage and low-power version of the popular 555 timer. Its nominal consumption is 60 µA at 3 V and it yields a stable frequency output.

Table 2.1 summarizes the current consumption of the different electronic modules integrated in the tag. The energy harvesting needed to feed the entire system is in the order of 1 mA at 3 V.

Element	Consumption (μA)
NFC (*M24*)	400
μC *(ATtiny85)*	300
Humidity *(HIH5030)*	200
Temperature *(LM75)*	280
Timer (*ICM555*)	60

Table 2.1. *Current consumption of the tag elements*

2.3.2. *Volumetric water content*

Following the discussion of section 2.2, there are several methods for measuring the Volumetric Water Content of the soil. The approach presented in this chapter is based on an interdigital capacitor, which is buried in the soil. The dielectric permittivity of the soil is highly dependent on changes in the Volumetric Water Content. Consequently, it is possible to measure the changes in dielectric permittivity as long as the strip-lines of the interdigital capacitor are kept in contact with the soil, and then retrieve the soil's VWC.

2.3.2.1. *Capacitor*

The capacitive sensor is implemented using an interdigital capacitor (IDC) without a ground plane (see Figure 2.3). Printed circuit board technology is particularly advantageous in implementing this type of sensor. The capacitive sensor prototype has been fabricated using a standard FR4 substrate with 34 μm-thick copper metallization. An insulator layer covers the IDC to prevent the electrodes from being short-circuited in the presence of water. A solder mask coating or a thin protection layer of plastic (this latter is the used in the prototype) can be used for this purpose. Figure 2.3 shows the shape of an interdigital capacitor, as well as the different parameters that influence its capacitance, which are: length L, gap distance between two conductors g, finger width s and the number of fingers N.

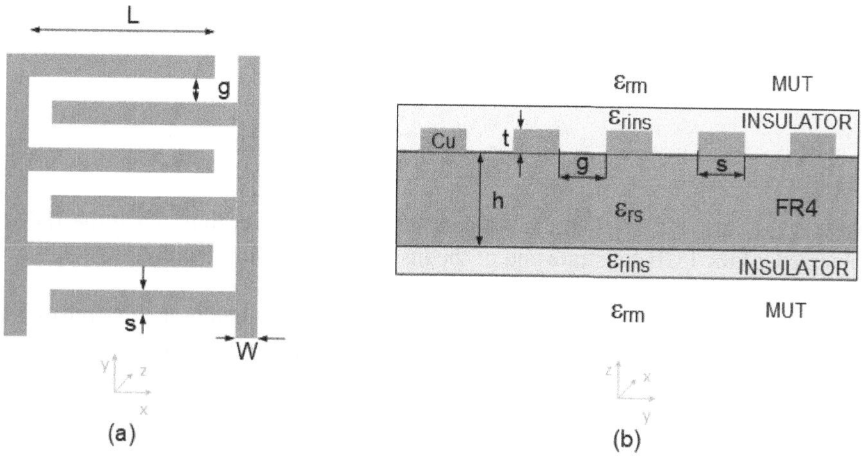

Figure 2.3. *(a) Top view of an interdigital capacitor. (b) Cross section of an interdigital capacitor*

Closed-form expressions for multilayer interdigital capacitors have been proposed in the literature (Gevorgian *et al.* 1996). The capacitance can be expressed as the sum of three contributions:

$$C = C_3 + C_n + C_{end} \quad [2.1]$$

where C_3, C_n and C_{end} represent the three-finger capacitance, the capacitance of the periodical (N-3) structure and a correction term for the fringing fields of the ends of the strips, respectively. These capacitances are proportional to the effective permittivity and the geometrical factors a_3, a_n and a_{end}, which are functions of the elliptic integrals of the first kind *K(k)*, resulting of the conforming mapping method (Svacina 1992):

$$C_3 = \varepsilon_{re3} a_3 L \quad [2.2]$$

$$C_n = \varepsilon_{rn} a_n (N-3) L \quad [2.3]$$

$$C_{end} = \varepsilon_{re,end} a_{end} \quad [2.4]$$

The effective relative permittivity ($\varepsilon_{re,i}$ with i = 3, n, end) can be calculated from the filling factor (q_i), which is the percentage of fields in each material for each case:

$$\varepsilon_{rei} = \varepsilon_{rm} + q_{1i}\frac{\varepsilon_{rins} - \varepsilon_{rm}}{2} + q_{2i}\frac{\varepsilon_{rS} - \varepsilon_{rins}}{2} + q_{3i}\frac{\varepsilon_{rins} - \varepsilon_{rm}}{2} \quad [2.5]$$

where ε_{rm}, ε_{rs} and ε_{rins} are the relative permittivities of the medium when the IDC is immersed, of the PCB substrate and of the insulator layer, respectively.

For long fingers (L >> s), the correction of fringing fields at the end of the fingers can be neglected (Gevorgian *et al.* 1996). Therefore, the capacitance *C* is mainly proportional to the length and the number of fingers (N). Substituting equations [2.2–2.4] into equation [2.1], the linear expression [2.6] for the capacity *C* of the IDC as a function of the medium permittivity expression can be found.

$$C = A + B\varepsilon_{rm} \quad [2.6]$$

where *B* depends on the geometry and *A* depends on the dielectric permittivity of the PCB substrate and insulator. The values *A* or *B* can be obtained using closed-form formulas (Gevorgian *et al.* 1996) or by simulation using a full-wave simulator.

The electromagnetic simulator Keysight-Momentum is used for the IDC design and evaluation. An IDC with N = 40, L = 30 mm, W = 0.8 mm, s = 0.8 mm, g = 0.5 mm, h = 0.8 mm, t = 34 μm and ε_{rs} = 4.7 is designed. The designed dimensions result in a value of capacitance that is easily readable and compatible with the minimum gap and trace width that can be achieved considering the available PCB manufacturing technology. Two prototypes with two insulator materials and thicknesses have been made. The first prototype uses an insulator (type 1) that consists of a coating of adhesive plastic (ε_{rins} = 2) with a thickness of 100 μm. In the second prototype, the insulator (type 2) is an acrylic coating spray (ε_{rins} = 2.7) with a thickness of about 15 μm.

Figure 2.4 shows the simulated IDC capacitance as a function of the relative permittivity of the material under test (MUT) where the IDC is immersed. The thickness of the MUT is assumed infinite. A linear dependence is observed when an ideal IDC without an insulator layer is analyzed. However, the insulator layer introduces a reduction to the sensitivity (slope as a function of relative permittivity) of the IDC as a function of the permittivity when values of the latter are higher than 10. In addition, the sensitivity is not constant, as predicted by the closed-form expressions derived from the partial capacitance method (Gevorgian *et al.* 1996).

The capacitance values are lower than those obtained in the case of having no insulator because the effective permittivity of the insulating layer and that of the MUT are less than ε_{rm}. The relationship between the effective permittivity and the permittivities of the insulator layer and the MUT is not linear, unlike what happens in multilayer microstrip lines (Svacina 1992).

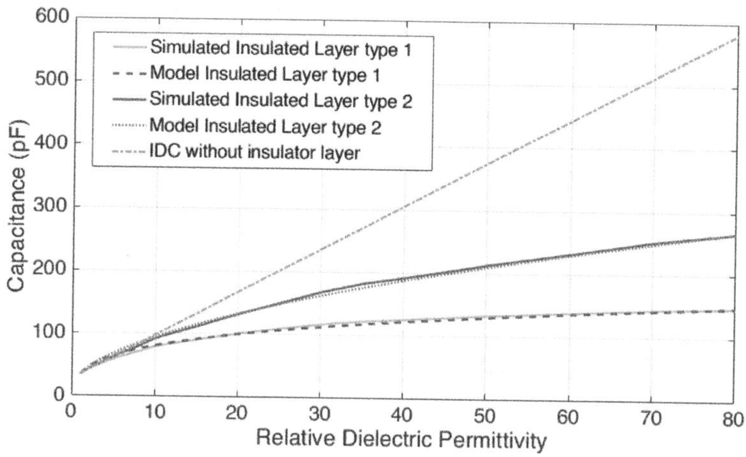

Figure 2.4. *Simulated and modeled capacitance as a function of the relative dielectric permittivity. Insulator layer type 1 is an adhesive plastic, and type 2 is acrylic coating spray*

Figure 2.5 shows a cross-section of the IDC sensor with the equivalent circuit for low (Figure 2.5(a)) and high (Figure 2.5(b)) permittivity of the MUT. In the first equivalent circuit, the capacitance due to the insulator layer (C_{INS}) is in parallel with the capacitance of the MUT (C_{MUT}), whereas in the second, C_{INS} is in series with C_{MUT}. The first equivalent circuit is used in the derivation of closed formulas from (Gevorgian *et al.* 1996). In the second equivalent circuit, when analyzing the fringing electric field path between the electrodes, it makes more sense for the capacitances of the insulation layer C_{INS} and of the MUT C_{MUT} to be connected in series for high dielectric constant values of the MUT (Kim 2008). Therefore, it is expected that the effective permittivity and capacitance will be smaller in the second equivalent circuit. Simulated values of C in Figure 2.4 agree with the measured capacitance when the IDC is in the air (air being the insulator layer, with $\varepsilon_{rm} = 1$), C = 23 pF. The C values also agree when the IDC is immersed in water ($\varepsilon_{rm} \approx 80$), C = 140 pF for the first prototype (insulator type 1) and C = 270 pF for insulator type 2. A compact expression is proposed in order to interpolate the capacitance values:

$$C = C_1 + (C_2 - C_1)\frac{(\epsilon_{rm})^{\gamma} - (\epsilon_{rm1})^{\gamma}}{(\epsilon_{rm2})^{\gamma} - (\epsilon_{rm1})^{\gamma}} \qquad [2.7]$$

where C_1 and C_2 are the capacitance for two known permittivities, $\varepsilon_{rm1} = 1$ (air case) and $\varepsilon_{rm2} = 80$ (water case), respectively. The exponent γ is a fitting parameter. A value of $\gamma = 0.21$ is obtained for the IDC of Figure 2.4.

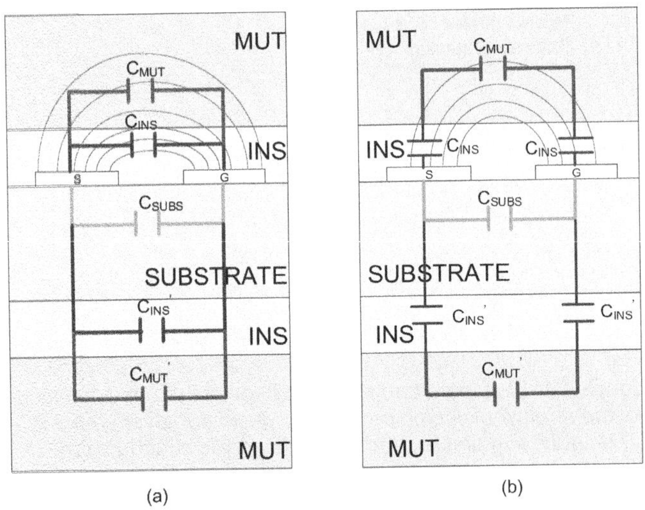

Figure 2.5. *Cross-section view of the IDC sensor with its superimposed equivalent circuit (a) for low permittivity of the MUT used in partial capacitance technique (Gevorgian et al. 1996) and (b) for the proposed equivalent circuit for high permittivity of the MUT (Kim 2008)*

2.3.2.2. *Capacitance measurement*

Reducing the clock frequency of the microcontroller to decrease its power consumption implies some restrictions. In order to minimize the number of tag components and, consequently, reduce its cost and complexity, the microcontroller's clock frequency is chosen using its own internal oscillator. Therefore, the resolution of the capacitance measurement based on the discharge time of the capacitor is not enough for small capacitances since, when it is in the order of pF, large time constants are expected. In an NFC communication, power is only available when the reader is close to the tag. Consequently, a fast measurement method must be applied. A similar problem arises when an oscillator whose frequency depends on the capacitance is used. In this method, the microcontroller works as a frequency counter and the resolution depends on the maximum clock frequency. Taking all these considerations into account, the methodology proposed here to measure the capacitance is shown in Figure 2.6. A custom frequency-to-voltage converter has been designed because the available commercial IC converters require higher currents than are available in this system. A low-power timer is used to generate an AC signal, which is filtered by a low-pass filter (RC filter), whose cut-off frequency depends on the capacitance under test. The AC amplitude at the output of the filter is measured using a half-wave rectifier that operates as detector. The DC voltage filtered at the output of the detector is measured with the internal analog-to-digital converter of the microcontroller. Although the output voltage depends on the power supply (V_{DD}), thanks to the fact that the microcontroller uses that voltage as a reference, this problem is avoided.

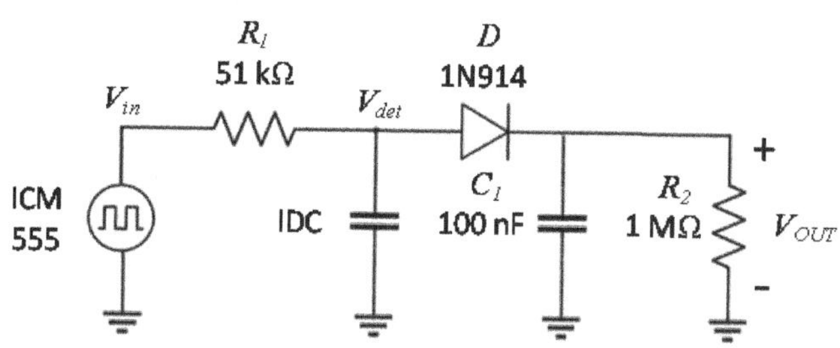

Figure 2.6. *Circuit to obtain a voltage proportional the interdigital capacitor (IDC) value*

The proposed circuit consists of a square wave generator of f = 380 kHz ($\omega = 2\pi f$) based on the 555 IC connected to a low-pass filter composed of a resistor of 10 kΩ (R_1) and the interdigital capacitor (IDC) whose capacitance value C varies in the range 55–270 pF. The resulting signal (V_{det}, which presents a triangular waveform) is rectified by a diode (D). The cathode of the diode is connected to a capacitor (100 nF, C_1) and a shunt resistor (1 MΩ, R_2). Hence, at the output of the envelope detector, V_{OUT} is proportional to the value of the IDC.

The waveform of the signal generated by the 555 timer (V_{in}) is a square wave with a nearly ideal 50% duty cycle. This periodic signal can be expanded using the Fourier series. The higher order harmonics are filtered; therefore, only the first harmonic is considered. The input signal at the filter can then be expressed as:

$$V_{in}(t) = \frac{V_{DD}}{2} + \frac{V_{DD}}{\pi}\cos(\omega t) \qquad [2.8]$$

Considering $H(j\omega)$ as the transfer function of the low-pass filter defined as:

$$H(j\omega) = \frac{V_{det}}{V_{in}} = \frac{1}{1 + j\omega R_1 C} \quad [2.9]$$

The voltage at its output is given by:

$$V_{det}(t) = \frac{V_{DD}}{2} + \frac{V_{DD}}{\pi} \lfloor H(j\omega) \rfloor \cos(\omega t + \varphi) \quad [2.10]$$

where φ is the phase of the transfer function. Therefore, the output of the half-wave detector is proportional to the average amplitude:

$$V_{OUT} = \frac{V_{DD}}{2} - V_f + k\lfloor H(j\omega) \rfloor V_{DD} \approx \frac{V_{DD}}{2} - V_f + \frac{kV_{DD}}{\omega R_1} \frac{1}{C} \quad [2.11]$$

where k is a factor that depends on the waveform at the output of the filter (V_{det}, nearly triangular), and V_f is an offset correction that considers the forward voltage of the diode. If the frequency of the oscillator is chosen to be higher than the cut-off frequency of the filter $f >> 1/(2\pi R_1 C)$, the signal at the output of the detector (V_{OUT}) is approximately proportional to the inverse of the capacitance C. Figure 2.7 compares the model given in equation [2.11], the simulation of the circuit in Figure 2.6 considering a conventional silicon diode (1N914), and the measurements read by the NFC reader. Good agreement has been obtained in the results, with $k = 0.5$ (an almost triangular waveform was considered at the filter output) and a forward voltage correction $V_f = 0.35$ V, obtained empirically. Different values of 0603 Surface Mount (SMD) capacitors have been mounted to emulate the IDC. Measurements have been made simultaneously with a multimeter (Agilent 34450) and a microcontroller that is powered by energy harvesting coming from a smartphone.

Performing a sensitivity analysis from the slope of the voltage as a function of the capacitance, the average error obtained from the 10-bits ADC of the microcontroller is shown in Figure 2.8 and is below 2%, considering the capacitance range (55–270 pF) from the capacitive sensor.

Figure 2.9 shows the prototype manufactured on an FR4 PCB. The parts of the design are pointed out: NFC antenna, the chipset with a connector that gives access to the microcontroller in order to program it, temperature and humidity sensors, and the interdigital capacitor which will be buried into the soil.

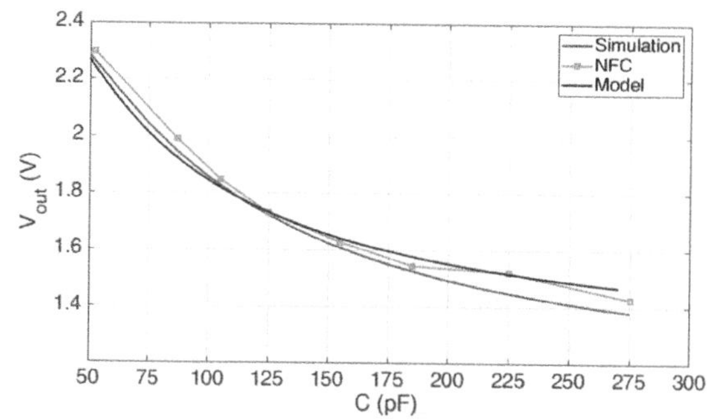

Figure 2.7. *Output voltage of the circuit to obtain a voltage proportional the interdigital capacitor (IDC) as a function of its capacity C. Comparison between the model proposed equation [2.11], the simulation and the NFC read using a smartphone*

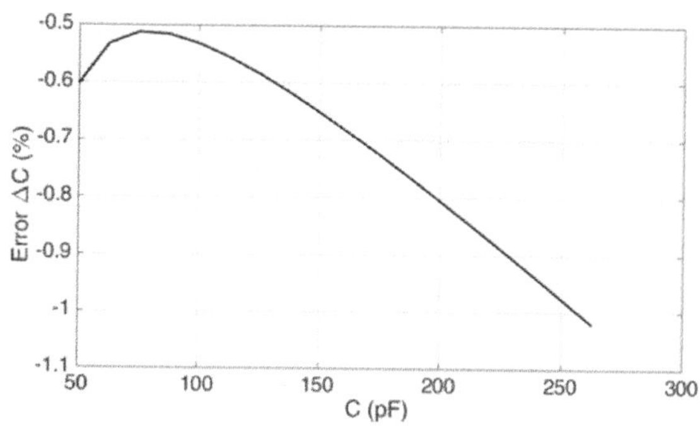

Figure 2.8. *Average error in the measurement of the capacitance from the analysis of sensitivity*

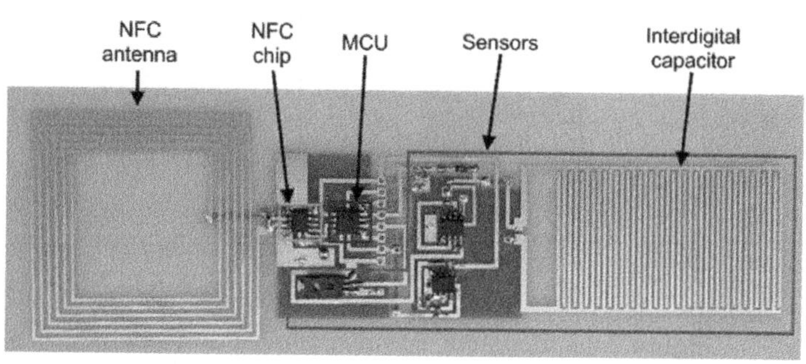

Figure 2.9. *Designed PCB prototype with the antenna, the IDC and the chipset*

2.3.3. *Software*

In order to read, display and save the sensed data, it is necessary to program both sides: the reader and the tag. Figure 2.10 shows the flow diagram of the system operation. Once the NFC tag IC receives a magnetic field higher than H_{min}, it supplies the rectified voltage to the tag electronics and activates the software programmed on the microcontroller, which gets the data from the sensors, processes it and writes it to the NFC IC memory in NDEF format. This NDEF message is retrieved and processed by the reader software.

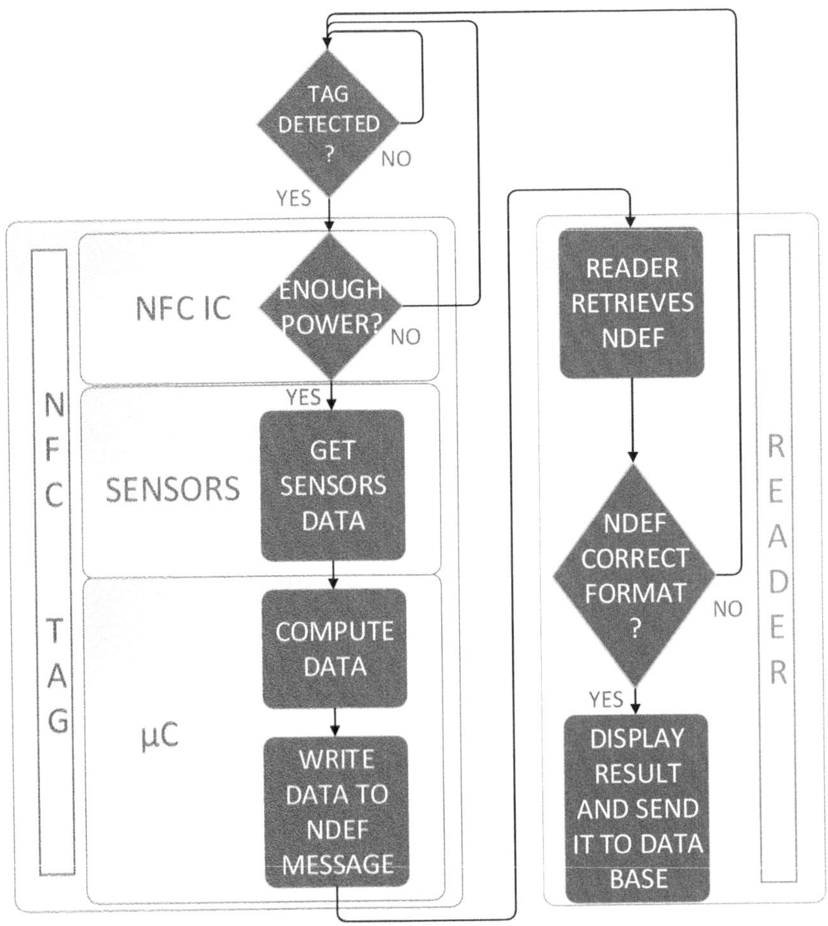

Figure 2.10. *Operation flow diagram of the system, indicating the tag on the left and the reader on the right*

2.3.3.1. *Tag side*

To compute and store the sensors' data, it is necessary to program the tag's microcontroller. Figure 2.11 shows the flow diagram of the tag operation. Once the tag is powered up by the harvested energy from the RF signal, the sensors and the electronics are powered. The first code to run will be dedicated to configuring the NFC IC energy harvesting options, which affects two registers of the system area. The first is at the I^2C location 2336, and contains the *EH_enable* bit that must be set to 1 to activate the energy harvesting option. However, this register is volatile, and it will be updated every time the IC is powered up according to the second register. The second register includes options regarding the energy harvesting, and can be accessed through the I^2C address 2320. In this register the *EH_mode* is configured, which indicates whether the energy harvesting mode is enabled by default or not. The same register contains 2 bits (*EH_cfg0* and *EH_cfg1*) used to choose the sink current level. For this application the maximum sink current is selected, which requires a minimum magnetic field strength (H_{min}) of 3.5 A/m, providing a current of up to 6 mA. The configuration of these registers can be done via I^2C from the microcontroller or by means of RF messages from the reader. The energy harvesting option is disabled by default. In order to set the desired energy harvesting setting, the tag must be externally powered. The microcontroller will program the registers. Alternatively, these registers can be configured by means of RF commands from the smartphone using an app provided by the manufacturer. Once the configuration byte is written, it will remain so unless it is overwritten. Nonetheless, the register is configured each time to ensure that the expected sink configuration is defined.

Timing is an important factor to take into account in NFC systems, and delays on the code must be considered. Thus, after writing the configuration registers, the microcontroller's firmware will wait 50 milliseconds to provide time to the sensors in order to avoid erroneous readings. After that, the microcontroller calculates the average of five readings from the 10-bit ADC, which is connected to the IDC (V_{OUT} in Figure 2.6) and calculates the VWC with the calibration curve that will be explained in section 2.4.1. The relative humidity is gauged with the average of five readings of the output voltage of the sensor and applying the equation provided by the sensor's manufacturer on its datasheet. The last parameter that is measured is the temperature, which is obtained from the digital sensor LM75A. To get the value, the microcontroller has been programmed to start a transmission via I^2C bus to the address of the sensor (configured by hardware) requesting the register in which the temperature is stored.

Once the data of the three sensors has been computed, the values are formatted to keep a fixed length and arranged in NDEF format to be sent to the NFC IC through the I^2C bus. The NDEF message is stored into the EEPROM of the NFC chip and it will be used to modulate the backscattered signal which will be received, demodulated and interpreted by the reader.

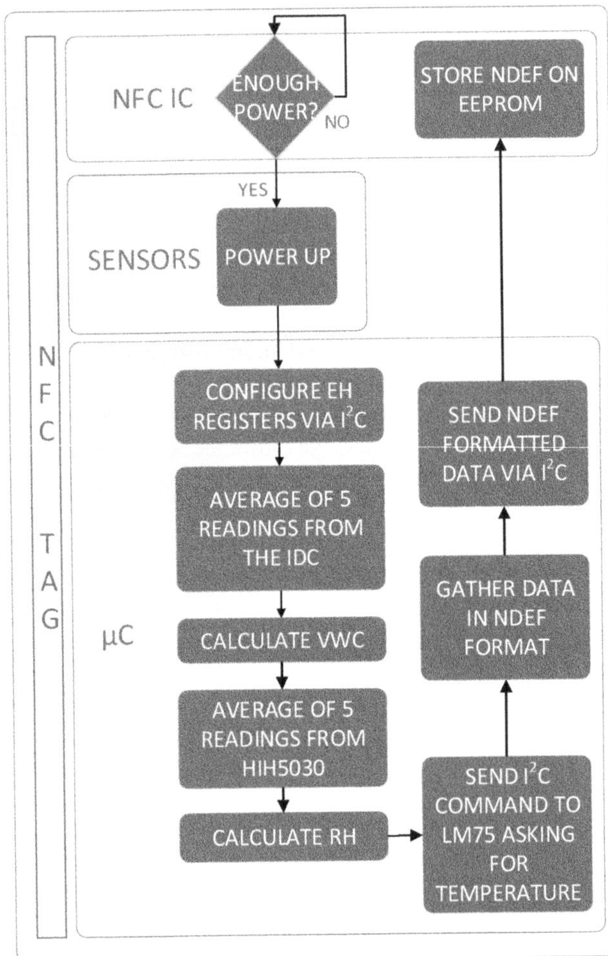

Figure 2.11. *Flow diagram of the tag firmware*

The main problem of the system occurs when the magnetic field received at the NFC IC from the reader is not enough to supply the rectified voltage to the tag's circuitry yet it is enough to read the tag. In that case, the reader will retrieve the last stored message on the memory instead of performing a new reading. This drawback can be solved by applying some type of timestamp or control bit on the NDEF message.

2.3.3.2. *Reader side*

An application that runs on an Android smartphone (shown in Figure 2.12) has been developed as proof of concept. Beyond the sensed data, the system makes it possible to send more information to the reader, such as a link to a website, or it can make use of the user identifier (UID) of each tag to give additional information on that specific product.

a) b)

Figure 2.12. *Image of the developed app graphic interface which uses the tag UID to retrieve the information of the plant (a) and the flowerpot with the device on it being powered and read from a smartphone (b)*

2.4. Experimental results

2.4.1. *Calibration*

An experimental setup and a calibration procedure are proposed in order to check the operation of the sensor. To carry out the experiments, samples of different types of soil used in greenhouses are taken, with different volumetric water content, ranging from dry to water saturation. The interdigital capacitor IDC is connected to a vector network analyzer (VNA), which is used to measure the capacitance at 1 MHz. Figure 2.13 shows both the measured capacitance as a function of the volumetric water content (blue squares) and the two-point α-mixing model [2.12] (blue line). A capacitance of 65 pF is obtained when the soil is dry, which is considerably higher than when the sensor is in the air (around 55 pF). Figure 2.13

also shows the soil permittivity obtained from the relationship between the capacitance and the permittivity given by equation [2.7].

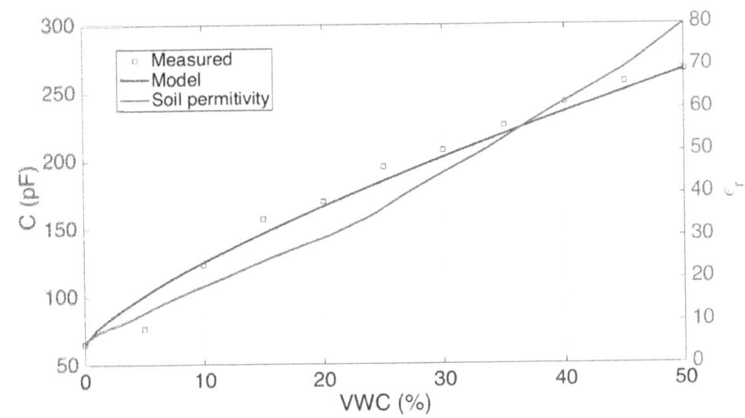

Figure 2.13. *Measured capacitance (blue squares), model [2.12] (blue line) and soil permittivity (orange line) as a function of the volumetric water content*

In order to achieve high accuracy, the sensor must be calibrated, because the relation between the capacitance and the volumetric water content depends on the soil composition (Sakaki *et al.* 2011). In addition, as is shown in Figure 2.4, the capacitance depends on the coating thickness and the coating material. To perform the calibration, the two-point α-mixing model is fitted (Sakaki *et al.* 2011) and the volumetric water content is modeled as:

$$VWC(\%) = \frac{X^{\alpha} - X_{sat}^{\alpha}}{X_{dry}^{\alpha} - X_{sat}^{\alpha}}\phi \qquad [2.12]$$

where X is a variable that depends on the sensor measurement, α is a fitting parameter that depends on the sensor, and ϕ is the porosity of the medium. The sub-indexes *dry* and *sat* correspond to the sensor output values when the soil is dry and water-saturated, respectively. The fitting parameter α depends on the sensor and is obtained from the slope in logarithmic scale:

$$\alpha = \frac{ln\left(1 - \frac{VWC}{\phi}\right)}{\ln(X)} \qquad [2.13]$$

Considering that the proposed circuit has an output voltage proportional to the VWC, it is possible to define X as the normalized measured voltage (Figure 2.6):

$$X = \frac{V_{out} - V_{out,min}}{V_{out,max} - V_{out,min}} \qquad [2.14]$$

where $V_{out,min}$ and $V_{out,max}$ are the minimum (in water) and the maximum (in air) measured output voltages, respectively. Therefore, using this normalized variable X, $X_{dry} = 1$ and $X_{sat} = 0$. α and ϕ can be calculated from a linear fitting in logarithmic scale.

The values obtained for α and ϕ are 0.33 and 50%, respectively. Figure 2.14 shows the results of the volumetric water content as a function of the normalized voltage, comparing the model with the measured values.

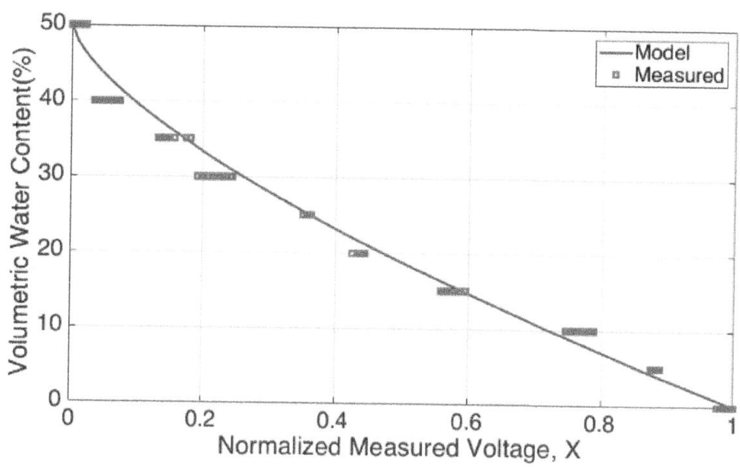

Figure 2.14. *Comparison of measured (Υ) and modeled volumetric water content as a function of the normalized voltage, X*

2.4.2. *Irrigation*

The measurement process was performed using 11 pots containing substrate of varying levels of wetness. The moisture percentage covers the range of dry soil up to 50% in steps of 5%. Thus, a homogeneous distribution of the water through the soil is assured. The sensor has been buried in each pot long enough to obtain 50 samples. The results are shown in Figure 2.15. The stability of the measured VWC can be observed until 45% saturation is reached. In order to simulate the drying process, the same procedure has been done from 50% to 0% VWC.

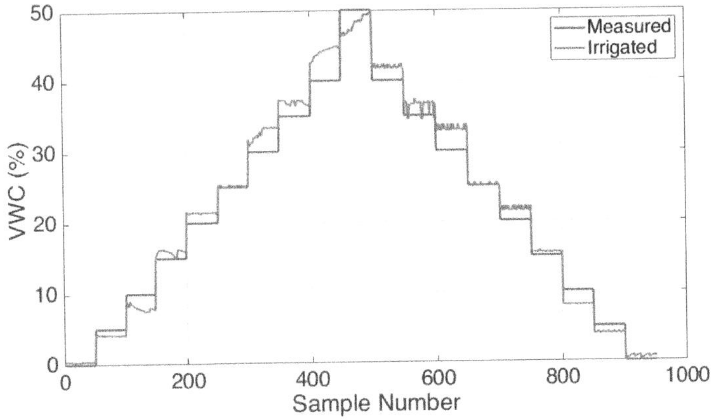

Figure 2.15. *Measured VWC compared to the irrigated water*

2.4.3. Comparison against commercial sensor

Figure 2.16 shows the behavior of the NFC sensor after watering a pot filled with soil with 150 ml of water. This pot had a capacity of 1 liter and initially contained about 13% of VWC. These measurements are compared with the results obtained with a commercial sensor (Decagon EC-5). As we can observe, the commercial sensor has a smoother response than the NFC sensor, which responds more quickly. The fast initial increase of VWC (at 5 min) is due to the presence of water on the sensor surface. After all the water has been distributed in the container, both sensors indicate approximately the same value.

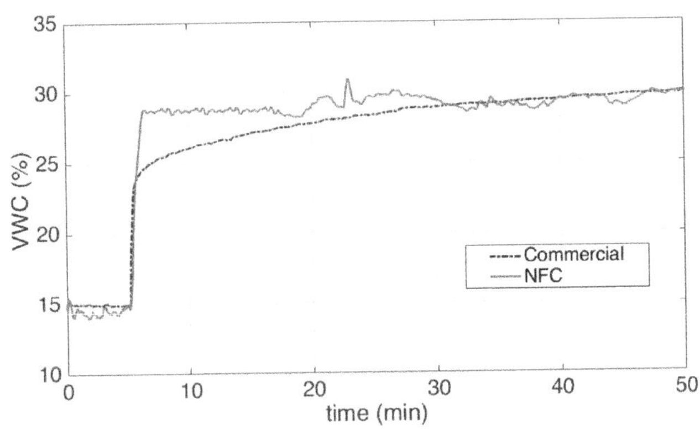

Figure 2.16. *Measurement of VWC after irrigation using the proposed NFC sensor and a commercial soil moisture sensor*

3

Case Study 2: Smart Diaper

3.1. Motivation

The main objective of this chapter is to present the application of NFC in diapers. To this end, we propose a smart diaper based on a wet detector, which is integrated in a battery-less NFC tag to be read from a smartphone.

Diapers are used to absorb moisture not only in babies but also in people like the elderly who have difficulty controlling their bladders or bowel movements or who are unable to get to a toilet quickly, such as bedridden people and those in a wheelchair. Since average life expectancy is increasing, the elderly population has grown, and the number of people who require extra care is increasing (WHO 1998; Wai *et al.* 2012). Sometimes, it is difficult to check the state of the diaper in babies and the elderly without removing their clothes. Nurses and other caregivers spend a lot of time performing this task. Therefore, optimizing this process can help reduce the cost of health services. Diapers must be changed to prevent skin rash caused by the skin being exposed for a long time to wetness and the high pH levels of urine and feces (Berg *et al.* 1994). It is therefore important to reduce the time that the moisture is in contact with the skin by regularly replacing the diaper (Adam 2008). Skin rash is often avoided by adding absorbent gels, but these can cause side effects such as irritation and allergies. Consequently, the use of smart diapers or normal diapers that include reusable tags may be of interest, and this chapter demonstrates that NFC could be a technology used for this purpose.

For a color version of all the figures in this chapter, see: www.iste.co.uk/girbau/sensors.zip.

Sections of this chapter have been taken from a previous publication by the authors of this book. from Lazaro, A., Boada, M., Villarino, R., Girbau, D. (2019). Battery-less smart diaper based on NFC technology. *IEEE Sensors Journal*, 19(22), 10848–10858. doi: 10.1109/JSEN.2019.2933289.

Several sensors have been proposed in the literature to detect the degree of moisture in diapers (Sidén *et al.* 2004; Yambem *et al.* 2008; Sheng *et al.* 2010; Nilsson *et al.* 2011; Sidén *et al.* 2011; Sajal *et al.* 2014; Simik *et al.* 2014; Karlsen and Dong 2015; McKnight *et al.* 2015; Ziai and Batchelor 2015). Diapers that incorporate these devices are often called smart diapers. Newer devices are capable of wirelessly transmitting the diaper's status to a mobile phone to set off an alarm (Ang *et al.* 2008; Wai *et al.* 2008, 2012; Khan 2018). However, these communication functionalities increase the cost of the diaper, and they require batteries that need to be replaced or recharged. These batteries can also be a source of contamination if they are not recycled.

This chapter presents a smart diaper that uses a battery-less NFC tag. In order to offer a low-cost solution, the tag is designed to be reusable and can be put in conventional commercial diapers. A smartphone with NFC capability is used to read the tag. The moisture detector is based on a capacitive sensor. This sensor detects the change in capacitance between two electrodes. The main characteristics of this NFC-based smart diaper are as follows:

1) It can detect urine without having to insert the tag into the diaper (simply by adhering it to the output layer).

2) It uses a capacitive detection method, which enables the volume of urine to be detected, rather than a resistive switch (Sidén *et al.* 2004; Sidén *et al.* 2011; Simik *et al.* 2014; McKnight *et al.* 2015), which enables the detection of only two states, that is, wet or dry.

3.2. Capacitive moisture detection

Figure 3.1 shows a block diagram of the NFC tag embedded in the smart diaper. The capacitive moisture detector consists of two parallel electrodes that sense the variation in capacitance between them due to the presence of urine. The electrodes are connected to a low-power microcontroller (Atmel ATtiny85) powered by the M24LR04E (M24) NFC IC from ST Microelectronics.

A prototype of a flexible tag has been designed on flexible Rogers Ultralam 3000 substrate ($\varepsilon_r = 3.14$, $\tan \delta = 0.0025$, thickness = 100 μm) (Rogers Corporation 2015). The antenna is a square spiral made up of six loops with an area of 25 × 25 mm. The width of the strips (w) and the spacing (s) between them are 0.5 mm. The antenna prototype was measured with a vector network analyzer, obtaining an inductance of 1.15 μH and a quality factor of 61 at 13.56 MHz in free-space (Q_2, equation [1.25]). However, the presence of the smartphone's metallic case (when brought closer to the tag to read it) reduces the inductance to 0.9 μH and the quality factor to 48, and also detunes the antenna (Boada *et al.* 2018b). The influence of the body is small because the diaper acts as a spacer. The resonance frequency of the tag was adjusted by adding a 100-pF capacitor in parallel to the input of the NFC IC.

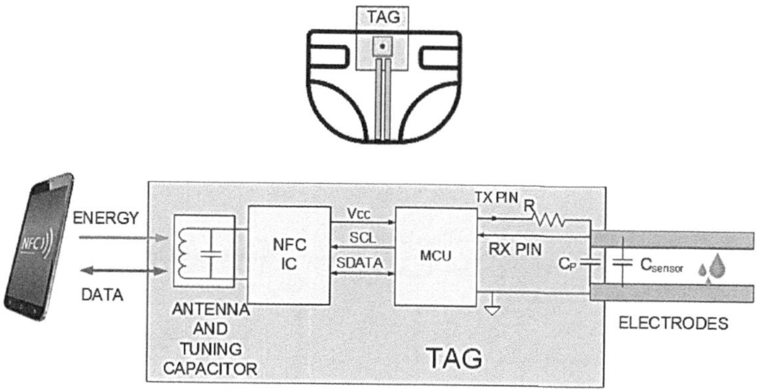

Figure 3.1. *Block diagram of the NFC tag for smart diapers*

A key commercial aspect is the cost of the prototype. For large quantities (>1000), the cost of the M24 and the ATtiny85 is around 0.3 € and 1 €, respectively. In commercial products, the Ultralam substrate can be replaced by another low-cost dielectric, such as polyamide. The total cost can therefore be below 1.5 €. The standard price of a package of adult diapers is roughly 15 € (with 10 units) and a package of baby diapers is roughly 20 € (100 units). Therefore, the commercial device, at least the electronic part, should be reusable. Consequently, the most interesting option is the electrode that can be printed with conductive ink on the surface of the diaper and the reusable tag that can be purchased as a separate element.

Figure 3.2 shows a cross-section of a typical diaper made up of several layers (Kosemund *et al.* 2009). The top layer is the one that is in contact with the skin. Its purpose is to transfer the fluids to the core while remaining soft and dry when in contact with the skin. Some diapers include a skin-care lotion that protects the skin from over-hydration and reduces irritation. The acquisition layer is placed between the top layer and the absorbent core. This intermediate layer channels the fluids away from the skin and spreads them through the diaper core for better absorption. The innermost layer of the diaper is the absorbent core. It generally consists of a blend of fluffy pulp made from cellulose and polyacrylate (polyacrylic acid sodium salt, $(C_2H_3NaO_2)_n$) granules. The cellulose material quickly absorbs and transfers the liquid to the polyacrylate superabsorbent material, where it is confined. This material is used because it can absorb and retain a large amount of liquid relative to its own mass. After absorbing the liquid, it solidifies into a gel. The diaper's outer layer is the back sheet, which is designed to be water-resistant to prevent liquids from seeping into clothing. This layer is made of polyethylene or a cloth-like film that lets in water vapor and air (but not liquids), thus reducing moisture and keeping the skin drier. As already explained, the NFC tag was designed on a flexible substrate to be attached to the back sheet on the outer surface, with the objective of avoiding any modification of commercial diapers.

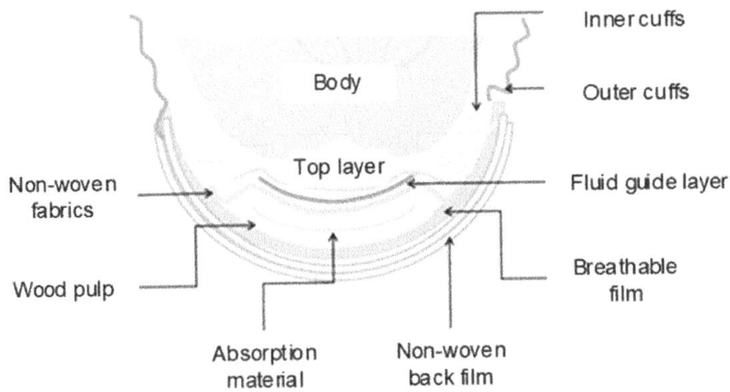

Figure 3.2. *Cross-section of a diaper*

Although the NFC IC can provide up to 5 mA at the maximum magnetic field, in this application, it is better to reduce current requirements of the tag in order to use a small loop antenna (especially in diapers for babies). To this end, the clock frequency is reduced to 1 MHz, achieving a maximum current consumption of 300 μA at 3 V. A capacitance measurement method that only requires the general-purpose input/output (GPIO) pins of the microcontroller and an external resistance R is used. Therefore, it does not add any active component that can increase power consumption. The method uses the charge time of the capacitive electrode, which is driven by a large resistor R (4 MΩ in the prototype) connected between a GPIO microcontroller digital output (TX PIN) and the electrode (see Figure 3.3). The microcontroller program toggles the TX PIN from a low to a high state, reads an input pin (RX PIN) and waits until its state is the same as that of the TX PIN. An integer variable is incremented inside a *while loop* until the RX pin state changes. The output of this counter is therefore proportional to the delay between the states. The delay Δt between the TX PIN change and the RX PIN change is determined by the RC time constant, where C is the total capacitance at the RX pin, including the electrodes and parasitic capacitances. The delay Δt can be computed as:

$$\Delta t = -R \cdot C \cdot ln\left(1 - \frac{V_t}{V_{DD}}\right) \qquad [3.1]$$

where V_t is the threshold voltage level that toggles a digital pin in the microcontroller and V_{DD} is the digital voltage power supply provided from the energy harvesting output of the NFC IC. If the RX PIN is configured as a digital input, this threshold corresponds to the logic threshold (usually $V_{DD}/2$ in CMOS). Another threshold voltage can be chosen if RX PIN is configured as an analog-to-digital converter input. The counter value (NFC reading) is proportional to the delay Δt, and therefore, to the capacitance equation [3.1]. The circuit operation is summarized in Figure 3.3(a), where the signals involved are plotted in the graph. The resolution is determined by the minimum time step that the microcontroller can measure, which is limited by the clock frequency and delays in the internal microcontroller. Although the proposed method is simple, it provides enough precision for this application. The accuracy involved in measuring C can therefore be increased by increasing the resistor value. This method is often used successfully in low-cost touch-button libraries in microcontrollers to avoid the need to add specialized integrated circuits based on measuring capacitance using the charge transfer method (Blake and Bible 2005; Gaitan-Pitre *et al.* 2009; Akkoç *et al.* 2016).

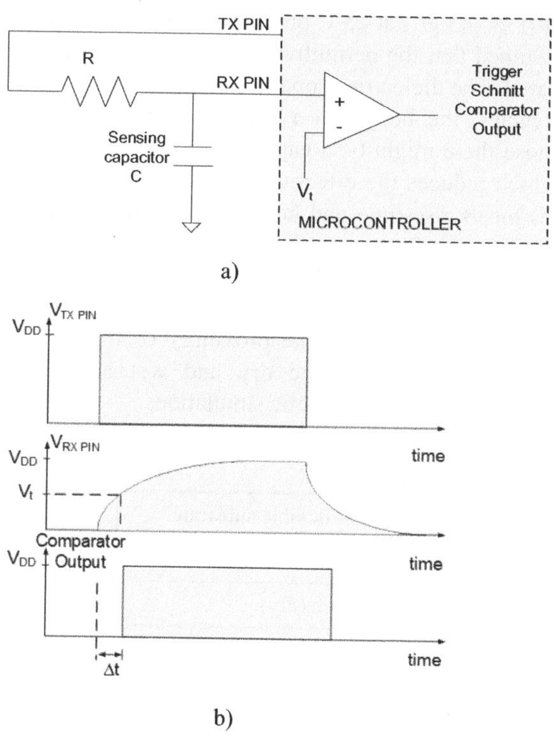

Figure 3.3. *(a) Scheme of the circuit used for measuring the capacitance. (b) Waveforms at TX PIN, RX PIN and at the output of the internal microcontroller comparator*

3.3. Capacitance simulation

Electromagnetic simulations to estimate the change in capacitance between the dry and the wet diapers are performed. The S parameters of two coplanar electrodes are simulated using the Keysight Momentum simulator. The capacitance between the electrodes is obtained from the imaginary part of the input admittance. Figure 3.4 shows the simplified cross-section of the diaper with the coplanar electrodes used for simulation. The thickness of the back sheet including the flexible PCB substrate is h_1. The permittivity of this layer (ε_{r1}) is assumed to be 3, which is the typical permittivity of polyamide. The permittivity of the absorption layer (ε_{r2}) is considered to be 2 when dry, which is the permittivity of the cellulose layers. The permittivity of urine is roughly $\varepsilon_r = 50$, and the conductivity is 1.75 at 1 MHz (Gabriel *et al.* 1996a). For laboratory experiments, urine was simulated with a solution of water with salt (NaCl). The relative permittivity of the solution is close to that of water ($\varepsilon_{r2} = 78$) and the conductivity can be adjusted by changing the salt concentration. In accordance with Stogryn's model (Stogryn 1971), a solution of normality 0.175 N (43 g/l) at 25°C produces conductivities that are close to those of urine. It is assumed that the permittivity is close to that of the liquid when the core layer is saturated. The dielectric properties of the body are frequency dependent and, at low frequencies, the body has a high permittivity. Permittivity is difficult to estimate because there might be a gap of air between the body and the top layer of the diaper, which reduces the effective permittivity. The maximum bandwidth of a pulse depends on its rise-time and here is considered to be 10 MHz, 10 times the microcontroller's clock frequency. The permittivity of the body (ε_{r3}) is considered equal to 170 in the simulations, which is the value of muscle permittivity at 10 MHz (Gabriel *et al.* 1996b). Note that this layer is included in the simulation for two purposes: to study the effect of the proximity of the body, and to estimate the threshold capacitance between the dry and wet states. Table 3.1 shows the parameters used in the electromagnetic simulations.

Parameter	Symbol	Value
Permittivity of back sheet and flexible substrate	ε_{r1}	3
Height of back sheet	h_1	200 μm
Permittivity of the core (dry)	ε_{r2}	2
Permittivity of the core (wet with water)	ε_{r2}	78
Height of the core	h_2	5 mm
Permittivity of the body	ε_{r3}	170

Table 3.1. *Parameters used in electromagnetic simulations*

The geometric parameters of the electrodes were chosen to achieve sufficient sensitivity. When the diaper is dry, the effective permittivity of the multilayer structure is expected to be low, and so a low capacitance value between electrodes is obtained. The effect of the high permittivity of the body should be small if h_2 is longer than the field penetration depth, T (see Figure 3.4). However, when the diaper is wet, the urine is trapped in the absorbent core layer and the capacitance increases several times. The field penetration depth T can be estimated using conformal mapping analysis (Chen *et al.* 2004):

$$T = \frac{S}{2}\sqrt{\left(1+\frac{W}{S/2}\right)^2 - 1} \approx W + S/2 \qquad [3.2]$$

The penetration depth is, therefore, a function of the width (W) and the spacing between electrodes (S). T must be of the order of the thickness h_2 in order to maximize the ratio between the wet and the dry capacitances.

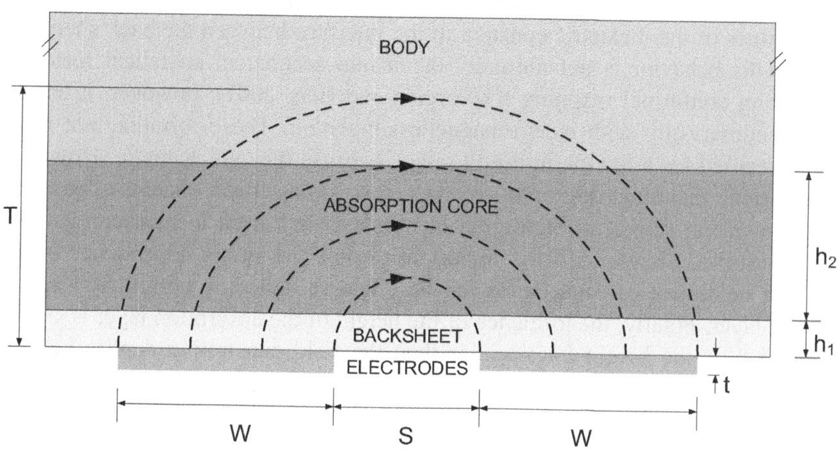

Figure 3.4. *Cross-section used for capacitance simulation showing the electric field distribution inside the diaper for low permittivity*

Figures 3.5–3.8 show the simulated results. To avoid excessive bending of the electrodes and improve contact with the diaper, the maximum width of the electrodes is considered to be roughly one quarter of that of the diaper (2–2.5 cm). Two cases are considered in all figures: (1) the diaper is in contact with the body and

(2) the diaper is in the air without contact with the body. The variation in the capacitance per unit length of the electrode (in pF/mm) is analyzed for the dry and wet diaper states. Figure 3.5 shows that the change in capacitance increases as the width of the electrodes W increases. As expected, C increases because the area of the electrodes increases. The capacitance is not affected by the body for widths less than the height of the core layer (h_2 = 5mm). For greater widths, the field penetration depth is greater than h_2 and there is a dependence on the permittivity of the body. Figure 3.5(b) shows the increase in capacitance normalized with respect to the dry capacitance. When the absorption core layer is saturated with water ($\varepsilon_{r2} = 78$), the ratio of capacitances ΔC/C can reach values of 8. Figure 3.6 shows the capacitance as a function of the separation between electrodes S. As the separation increases, the capacitance decreases steadily. W is assumed to be 5 mm in this figure ($T<h_2$ with a small effect of the body). However, an increase in the relative capacitance occurs when S increases (Figure 3.6(b)). Figure 3.7 shows the relative capacitance as a function of the permittivity of the absorption layer. This permittivity is a function of the urine content, the pressure and the diffusion of the urine along the electrodes. Capacitance is expected to behave linearly with permittivity. However, this linear effect is observed only up to 10. This effect is due to the curvature of the field lines for high values of the dielectric constant in the interface between the back sheet and the core. This behavior is not normally taken into account in analytical formulae derived using conformal mapping (Gevorgian and Berg 2001); however, it can be computed numerically with electromagnetic simulation. These formulae are often used and verified for a low permittivity range, because they are accurate. However, this conformal mapping does not consider this strong field change. The same behavior has been found in interdigital electrodes when used in measurements of soil moisture (see Chapter 2). It is important to note the strong dependence of the normalized increment of capacitance on the effective dielectric permittivity of the absorption layer. Finally, the influence of the height of the absorption layer is shown in Figure 3.8. If the height h_2 is higher than the field penetration depth, then the capacitance remains constant. For lower heights, T increases. For thicker layers, the normalized increment is almost constant. The height of the absorption layer depends on the diaper manufacturer and plays an important role in the price of the diaper and in the maximum volume of liquid the diaper can absorb. The absorption material increases in volume when the trapped urine content increases. Therefore, the diaper is thicker when it is wet rather than when it is dry, when the cellulose layers are compressed.

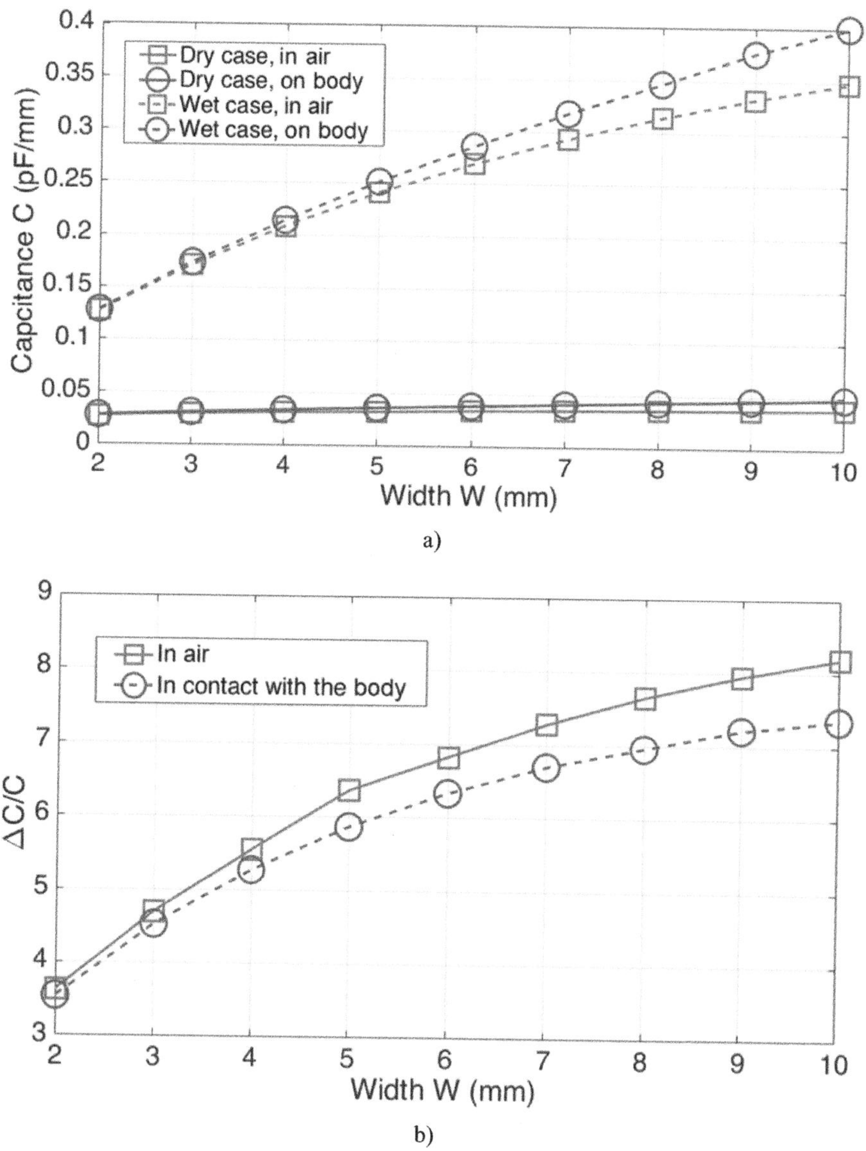

Figure 3.5. *Simulated capacitance per unit length (a) and normalized increment in capacitance (b) between the wet and dry cases as a function of the electrode width W, with the diaper in contact with the body and the diaper in the air (S = 1 mm, h_2 = 5 mm)*

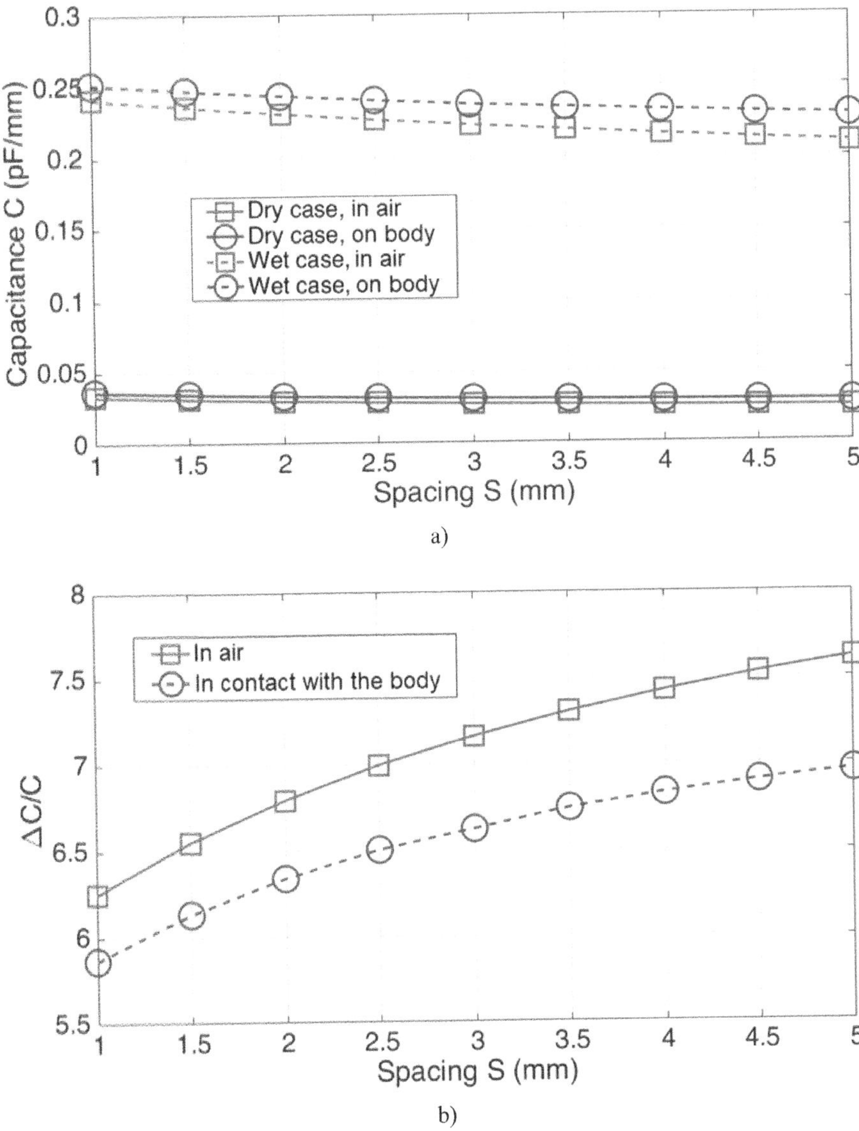

Figure 3.6. *Simulated capacitance per unit length (a) and normalized increment in capacitance (b) between the wet and dry cases as a function of the electrode spacing S, with the diaper in contact with the body and with the diaper in the air (W = 5 mm, h_2 = 5 mm)*

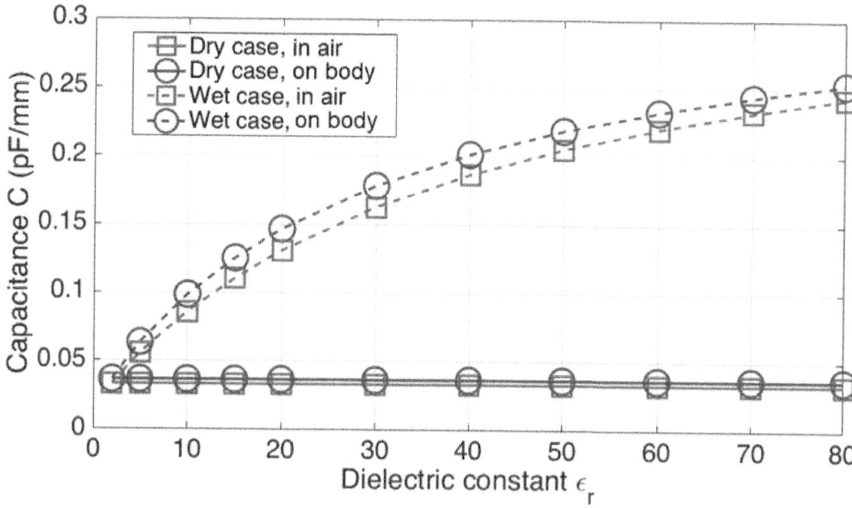

a)

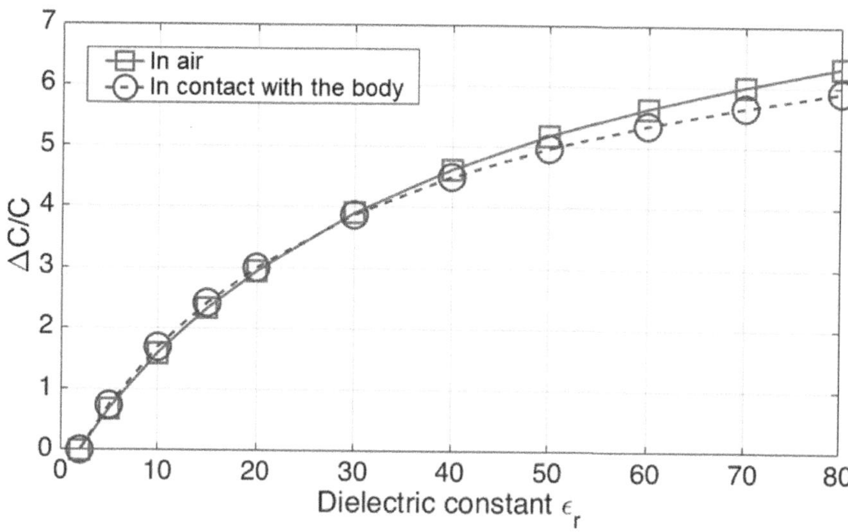

b)

Figure 3.7. *Simulated capacitance per unit length (a) and normalized increment in capacitance (b) between the wet and dry cases as a function of the effective permittivity of the absorption layer, with the diaper in contact with the body and with the diaper in the air (W = 5 mm, S = 1 mm, h_2 = 5 mm)*

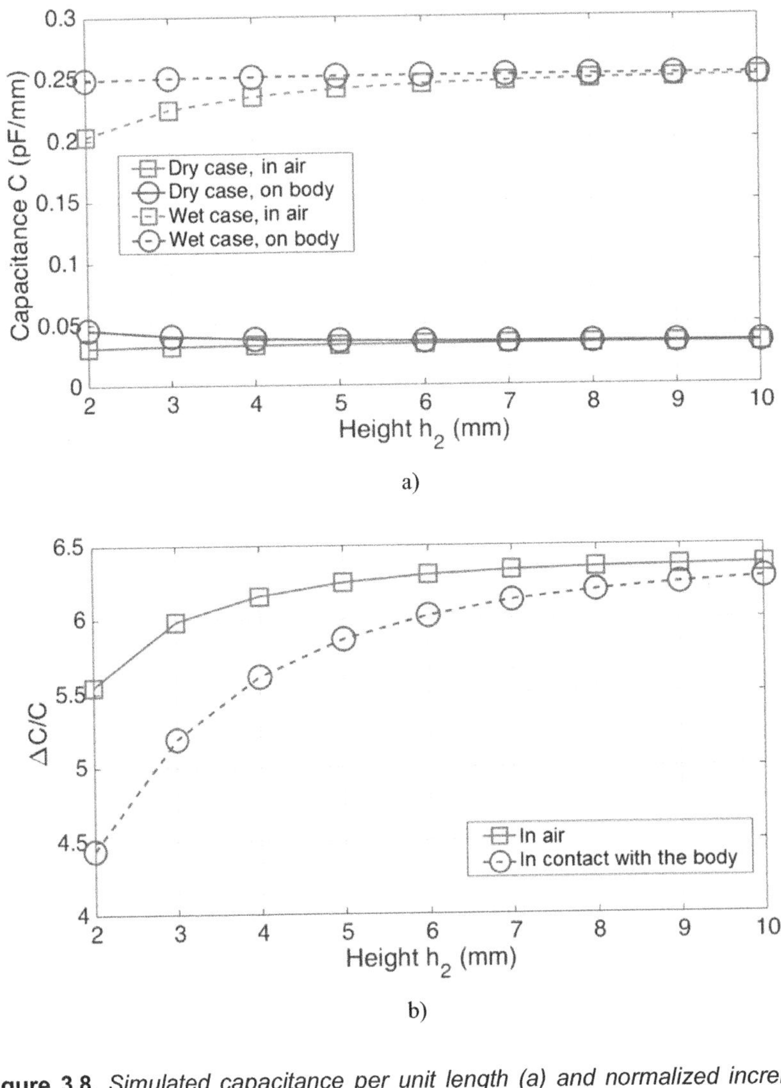

Figure 3.8. *Simulated capacitance per unit length (a) and normalized increment in capacitance (b) between the wet and dry cases as a function of the height of the absorption layer, with the diaper in contact with the body and with the diaper in the air (W = 5 mm, S = 1 mm)*

3.4. Experimental results

The initial measurements are devoted to checking the accuracy of the proposed capacitance measurement method. To this end, a set of measurements are conducted.

Figure 3.9 shows the count value measured by the tag when discrete surface mount device (SMD) capacitors are used instead of the electrodes. A resistor of 5 MΩ is used in these measurements. A linear regression is performed which clearly indicates that a linear behavior is obtained. The error between the model (line) and the measured values is under the tolerance of the capacitors (5%). The repeatability of the measurement was below 1 count or 0.18 pF. NFC measurements are performed by the means of two smartphones and the readout is identical in both cases. The offset error associated with parasitic capacitances and systematic errors is 23.2 pF. This accuracy is enough for the threshold detector.

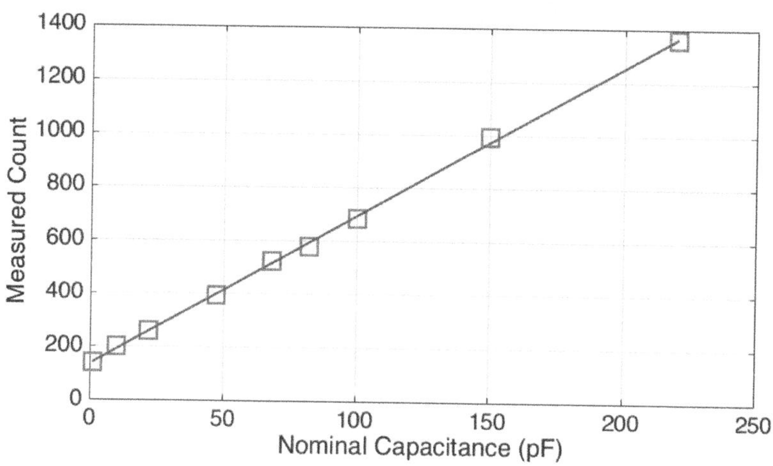

Figure 3.9. *Measured count with the NFC as a function of discrete SMD capacitors C*

Figure 3.10 shows a photograph of the tag, a photograph of the tag when adhered to a diaper and several mobile app screens designed to interact with the user. The experiments were performed with different diaper models, and a saline solution (0.175 M) was used to simulate urine.

Figure 3.11 shows the temporal response of the sensor when a volume of 120 cm^3 is injected uniformly with a pump to simulate urine discharge. An adult night diaper is chosen. It can be seen that the capacitance, and therefore the counter output (NFC reading), increases rapidly. However, there is a transitory period of roughly 100 seconds before the value stabilizes to the end value due to the diffusion of the liquid in the core. The rise time is of the order of 3.66 seconds, which demonstrates that the diaper can absorb urine flow rates of 33 ml/second. This is higher than the peak urine flow rate of roughly 27.3 ml/second for the average adult (Drach *et al.* 1979).

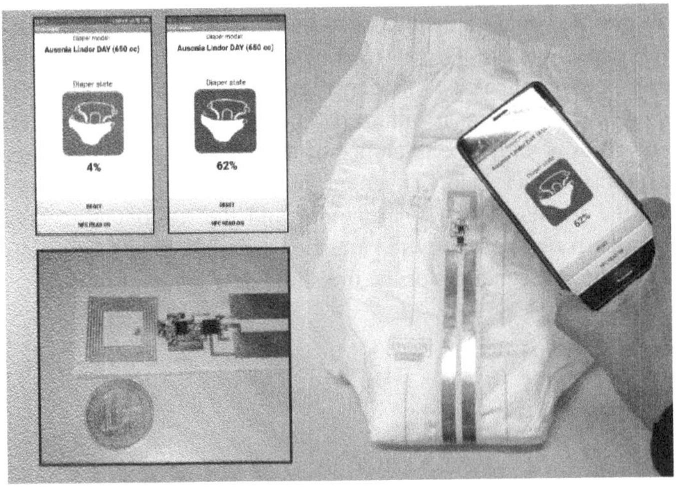

Figure 3.10. *Layout of the tag and the screen of the mobile app*

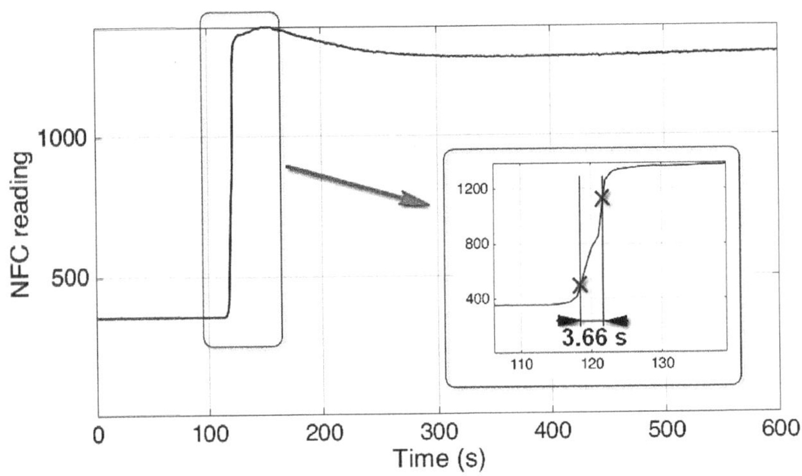

Figure 3.11. *Time-dependent sensor response to a 120 cm^3 urine volume injected into the diaper. The inset shows the detail of the rise time response between 10% and 90% of the end values*

Figure 3.12 shows the raw data output of the NFC sensor (output of the time counter) as a function of the volume of saline water. Two electrode widths are compared (W = 5 mm and W = 10 mm). The loop antenna is located in the belt region to make reading easier. In the experiments, the spacing (S) and the length (L)

of both electrodes were 5 mm and 20 cm, respectively. The increase in capacitance between the dry and the wet diapers was roughly 1.5 times higher when $W = 10$ mm than when $W = 5$ mm, which is in agreement with the simulations in Figure 3.5. Figure 3.13 compares the measurement of the NFC sensor and the capacitance of an LCR (L = inductance, C = capacitance, R = resistance) meter (Keysight U1733C) at 100 kHz. According to the manufacturer's datasheet, the accuracy of the LCR meter is 2%. The capacitance measurement error is therefore between 0.14 pF and 8 pF. A correlation coefficient of 99.5% is obtained between the two measurements. The small discrepancies along the line are mainly due to the diffusion of water in the absorption layer. This effect is especially visible when the liquid concentration is low, and the liquid does not cover all the electrodes' area. To mitigate this effect, the measurements are taken 5 minutes after the liquid is poured onto the diaper, which ensures the diffusion of the liquid within it. A variation of the measured capacitance in the range between 7 pF and 395 pF is observed, for water volumes between 0 and 400 cm^3 when $W = 10$ mm. The simulated and experimental results show that wider electrodes are preferable because they provide higher capacitance values and a higher capacitance ratio between saturated and dry states. In addition, it is also easier to cover the surface where the absorption layers are. Depending on the number of absorption layers in the core, on the superabsorbent polymer (SAP) mass ratio, on the pulp of cellulose used (Li *et al.* 2015), and on the expansion of the diaper in its wet state, the capacitance (and therefore the NFC reading) varies as a function of the model. Diapers designed to contain high volumes (up to 1,000 cm^3) can support several discharges during the night and contain several absorption layers. Diapers for babies can support up to 400 cm^3–500 cm^3 and their height is smaller than those designed for adults at night. However, when the volume reaches 200–250 cm^3, the urine covers all the area and the absorption layers start to expand, thus increasing the height of the diaper. The capacitance then tends to saturate, as we can see in Figure 3.8, where the capacitance remains almost constant when the height is greater than the penetration thickness. In the case of electrode 1, the resolution is roughly 0.16 cm^3 for the adult night diaper and 0.8 cm^3 for the baby's diaper. These resolutions are several times lower than they are for average urine discharge.

Figure 3.12 shows that the readings (and capacitance) increase steadily as the level of liquid increases until saturation is reached when the core is saturated, and the diaper is considered full. The capacitance or NFC reading can therefore be used to estimate the volume of liquid in the diaper. The following model is proposed to this end:

$$V(\%) = 100\frac{X^{\alpha} - X_{dry}^{\alpha}}{X_{sat}^{\alpha} - X_{dry}^{\alpha}} \quad [3.3]$$

where V is the percentage of liquid volume in the diaper, X is the normalized NFC reading, and α is the slope in logarithmic scale. X_{dry} and X_{sat} are the normalized NFC readings for the dry and water-saturated cases, respectively. Note that parameters X_{dry} and X_{sat} depend on the diaper model, but they can be saved in the NFC message that is sent to the mobile.

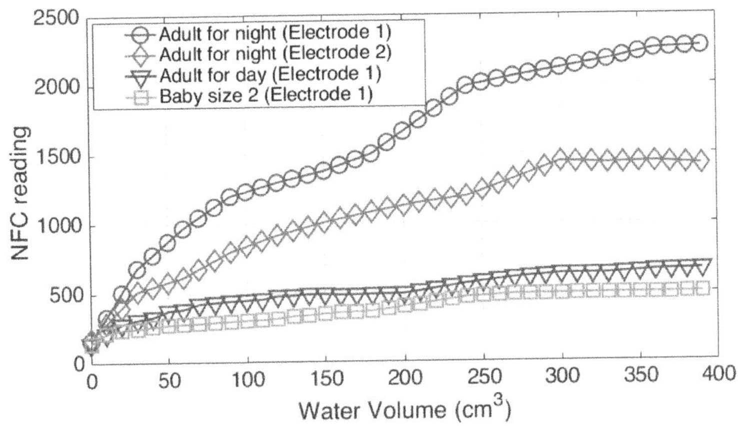

Figure 3.12. *NFC reading as a function of the level of saline water for two electrode widths (Electrode 1: W = 10 mm, S = 5 mm, L = 20 cm and Electrode 2: W = 5 mm, S = 5 mm, L = 20 cm) and different diapers (adult for night, adult for day and baby size 2)*

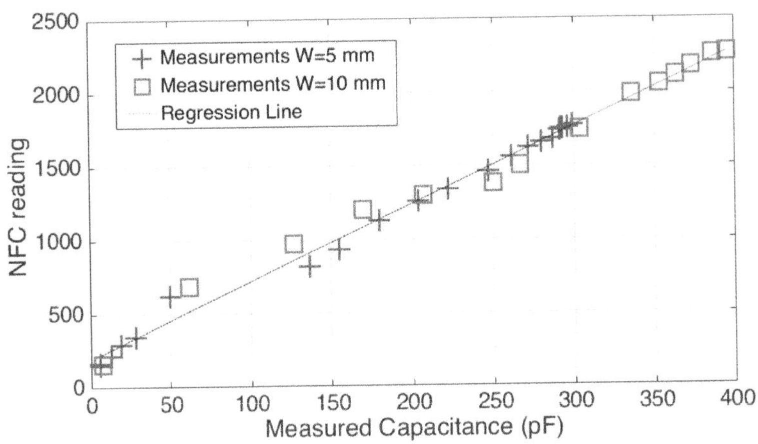

Figure 3.13. *Comparison of the capacitance between the electrodes and the NFC reading for two electrode widths (W = 5 mm and W = 10 mm). The length of the electrodes is L = 20 mm and the electrode spacing is S = 5 mm. A regression line is included. The correlation coefficient is 99.5%*

Figure 3.14 compares the proposed model [3.3] and the measurements for the diapers in Figure 3.12. A sound agreement is obtained for $\alpha = 2.5$. The correlation coefficient between the model given by equation [3.3] and the measured normalized volume is between 96% (baby diaper size 2) and 99% (adult day diaper). The standard deviation of the error in the estimation of the volume is between 3.1% (adult day diaper) and 10.3% (baby diaper size 2). Although, as can be seen in Figure 3.12, the dry and saturation values are a function of the diaper model, and these values can be stored by the diaper manufacturer in the EEPROM of the NFC IC and saved in the NFC message read on the smartphone app.

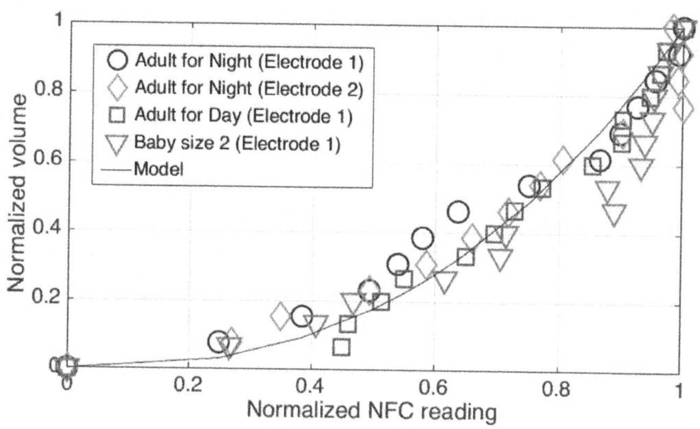

Figure 3.14. *Comparison of the models proposed for the normalized volume equation [3.3] as a function of the normalized NFC reading with the measurements for the diapers in Figure 3.12*

"Normal" urinary frequency depends on how much fluid is ingested per day, the composition of these fluids, and whether the person is being medicated. For most people, the normal number of times to urinate per day is 6–7 over a 24-hour period (Fischbach and Dunning 2009). Normal urine output depends on age (Drach *et al.* 1979; Van Hoeck *et al.* 2007; Fischbach and Dunning 2009). The normal output is 2–3 cm^3/kg/hour for neonates, 1–2 cm^3/kg/hour for infants, and 0.5–1 cm^3/kg/hour for adolescents and adults (Fischbach and Dunning 2009). Urine outputs below these limits are considered oliguria (Fischbach and Dunning 2009). For an adult who weighs 70 kg, a normal urine discharge is therefore between 120 cm^3 and 280 cm^3. As adult night diapers are designed for a maximum capacity of 1000 cm^3, they can support more than two discharges. Figure 3.12 shows that for volumes above 250–300 cm^3, the reading is almost constant, thus indicating a risk of overfilling. For a neonate whose weight is 4 kg, normal discharge is roughly 30–50 cm^3. The experimental results in Figure 3.12 show that these threshold volume values may be detectable.

To investigate the robustness of the readings in the presence of external materials below the diaper (chair, wheel chair, etc.), several experiments are conducted. The worst case is the presence of metal under the diaper (e.g. a metal chair), which can increase the capacitance between the electrodes. Figure 3.15 compares the NFC readings as a function of the low-permittivity spacer (cardboard, $\varepsilon_r \approx 1.8$), which simulates the thickness of clothes or chair cushion. An adult night diaper with electrode 1 was used in Figure 3.15, but similar conclusions have been found when other diaper models were used. As a reference, the measurement taken when the diaper was in contact with the body is also included. In the measurements in Figure 3.15, a phantom consisting of a thick block of foam weighting 4 kg was used to simulate body pressure. In agreement with simulated results, differences of roughly 3.5% were observed between the low permittivity phantom and the body. Looking at Figure 3.15, it can be concluded that a spacer of roughly 5 mm is sufficient to offset the effect of the metal. However, even when the metal is very close to the electrode, it is possible to determine whether the diaper is dry or wet.

On the contrary, maximum differences of 1.1% are observed between the diaper when flat and when it has a curvature radius of 8 cm. Therefore, if the electrode is attached correctly to the diaper surface, the effect of bending is not significant.

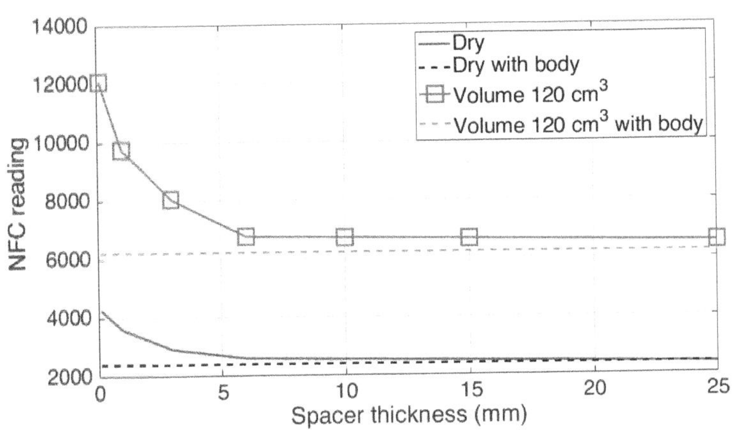

Figure 3.15. *Comparison of the NFC readings in terms of the distance to metal. The dashed lines show the values measured in contact with the body*

The repeatability of the measurements is illustrated in Figure 3.16. The histograms of 100 measurements for five diapers of the same model (adult night diaper) with the same volume of liquid (120 cm^3) are shown. The difference in the measurements for each diaper is under 0.5% and the maximum difference with respect to the mean is ±2%. These differences can be attributed to small differences due to the location of the electrode, the distribution of the liquid in the absorption layers and the surface roughness.

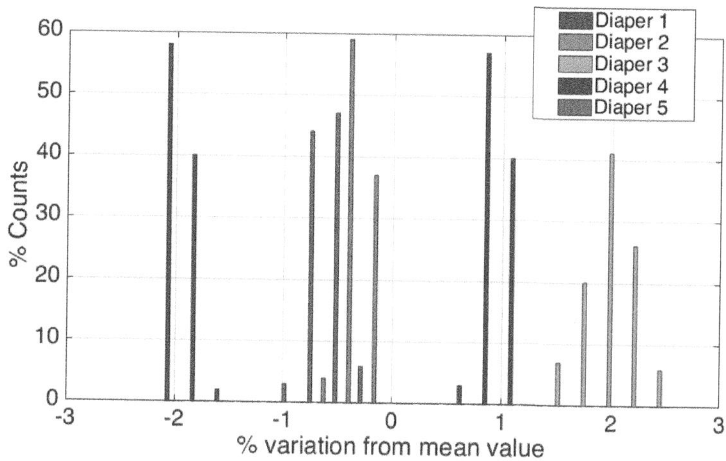

Figure 3.16. *Histogram of variation in NFC reading from the mean for five adult night diapers with a volume of 120 cm³*

3.5. Comparison with other technologies

Table 3.2 compares different passive (battery-less) and active smart diapers that use different technologies. Some techniques for sensing moisture in diapers are based on detecting changes in the color of a reactive in contact with urine. Such techniques are low-cost, but they cannot detect the state of the diaper under clothes. However, because of their simplicity, these technologies are commercially available in some diapers. Moreover, by using nitride, a protein and leukocyte reactive, they can detect the presence of certain infections based on a color change (Chen *et al.* 2014; Chen and Dong 2015). Other studies propose passive tags that are detuned in the presence of moisture. A UHF tag is used in Sajal *et al.* (2014), whereas an HF tag is used in Sidén *et al.* (2011) and Ziai and Batchelor (2015). The main drawback is that the read range is reduced for the wet state. This reduction may also be due to other reasons, such as misalignment between the reader and the tag antennas. Moreover, in Ziai and Batchelor (2015), the tag must be inserted in the core layer. In Sidén *et al.* (2011), an NFC tag with energy harvesting based on the SL13A IC from AMS is used, which performs sensing by means of a resistive switch. The main drawback is that the electrodes must either be inserted into the core layer (drilling the back sheet) or in contact with the skin. In both Sidén *et al.* (2011) and Ziai and Batchelor (2015), the commercial diapers must be modified and only a binary reading (wet or dry) can be obtained. In Yambem *et al.* (2008), the wet stage produces a shift in the resonance frequency of an LC resonator that is read with a non-standardized custom reader.

Year/reference	Sensor technology	Communication technology and comments
This chapter	Capacitive	NFC based on M24, estimates the volume of liquid, does not require modifying the diaper
Yambem *et al.* (2008)	LC resonance shift	Inductive coupling, only dry or wet states detection, modification on the diaper is required, custom reader
Sidén *et al.* (2011)	Resistive switch	NFC based on SL13A, only dry or wet states detection, modification of the diaper is required
Nilsson *et al.* (2011) and Sajal *et al.* (2014)	UHF antenna detuning	UHF RFID, measurement based on the horizontal and vertical polarization reading distances, only dry or wet states detection
Chen and Dong (2015) and Karlsen and Dong (2015)	Colorimetric	Not available, it cannot be used with clothes
McKnight *et al.* (2015)	Paper-based biopotential	BLE, requires external amplifiers and an impedance analyzer (AD5933)
Ziai and Batchelor (2015)	Resonance frequency detuning	Passive, HF at 13.56 MHz, only dry or wet states detection
Sidén *et al.* (2004)	Resistive switch	TX at ISM 13.56 MHz, activated by the sensor switch and custom reader
Simik *et al.* (2014)	Resistive switch	GSM, only dry or wet states detection, modification of the diaper is required
Khan (2018)	Temperature	BLE, requires continuous monitoring of the temperature
Rahman *et al.* (2017)	Resistive switch	BLE, only dry or wet states detection, modification on the diaper is required
Yu *et al.* (2016)	Resistive switch and colorimetric nitride sensor	BLE, urine-activated urinary nitride sensor

Table 3.2. *Comparison of smart diaper devices*

Systems based on active devices have longer read ranges than passive alternatives. However, a battery is needed and therefore the cost is higher. The first devices reported in the literature, such as Sidén *et al.* (2004) and Simik *et al.* (2014), use a custom TX transmitter. However, these communication systems usually consume more power, and battery life is limited. With the appearance of smartphones with Bluetooth Low Energy (BLE), several approaches based on this technology have recently been developed (Yu *et al.* 2016; Rahman *et al.* 2017; Khan 2018). Most of these approaches are based on the resistive switch, so an electrode must be in contact with the core of the diaper. In Khan (2018), a method based on measuring the external temperature of the diaper was proposed. This non-invasive method can detect a rise in the temperature of the outer sheet layer as a result of urine accumulation at a temperature above ambient. Unfortunately, this detection method cannot be applied to battery-less devices such as the one proposed in this chapter since it requires the temperature to be monitored periodically. However, it can be used if a battery is included, like in a data logger. Unlike the method proposed in this chapter, the smart diapers reported in the literature only provide information if the diaper is dry or wet and do not estimate the volume of liquid that can be used to determine certain pathologies associated with oliguria (Prowle *et al.* 2011). Unlike other HF RFID solutions for smart diapers such as Yambem *et al.* (2008) and Ziai and Batchelor (2015), the proposed method is compatible with commercial smartphones and does not require a specific reader.

4

Case Study 3: NFC Sensor for pH Monitoring

4.1. Motivation

Color analysis by microfluidic paper-based devices (Martinez *et al.* 2010) is widespread in applications such as medical devices (Oncescu *et al.* 2013; Sicard *et al.* 2015) and pH measurements (Vuong *et al.* 2010; Loh *et al.* 2011; Kim *et al.* 2017). The main advantages of these devices are that they are inexpensive, do not require an energy source, allow easy transport of fluids and are easy to use. However, since diagnosis depends on optical measurements, accuracy is poor especially when classification between different levels is required, as for instance in pH measurements.

pH measurement has a wide range of applications. It is used, for example, to check the water condition of swimming pools, tap water and aquariums. It is also used in medical applications such as sweat or saliva analysis and skin control in wounds, where it can predict an infection before it becomes visible. However, electronic pH probes are expensive and require continuous calibration. In consequence, the cost is prohibitive for home applications. In medical applications such as urine tests (Hashimoto *et al.* 2017), the probes must be sterilized each time they are used in order to avoid contamination between samples. For this reason, paper-based color strips have been extended to become the simplest, cheapest and quickest pH indicator. Since the integration of high-quality cameras in smartphones,

---

For a color version of all the figures in this chapter, see: www.iste.co.uk/girbau/sensors.zip.

Sections of this chapter have been taken from a previous publication by the authors of this book. Boada, M., Lazaro, A., Villarino, R., Girbau, D. (2019). Battery-less NFC sensor for pH monitoring. *IEEE Access*, 7, 33226–33239. doi: 10.1109/ACCESS.2019.2904109.

several methods for colorimeter measurement have been proposed (Vuong *et al.* 2010; Loh *et al.* 2011; Sicard *et al.* 2015; Kim *et al.* 2017). The main advantages of pH measurement using smartphone cameras are that they are easily available and portable. However, they also have notable drawbacks. For example, the readout depends on ambient illumination and camera type and requires special image processing techniques; therefore, there are variations between smartphone models and the programs used to process the images. To solve the problems related to variation in ambient light conditions, some studies propose the use of an enclosure to keep the sample in the dark and only be illuminated by the LED camera of the smartphone (Loh *et al.* 2011; Oncescu *et al.* 2013; Kim *et al.* 2017). This method allows the reading to be independent of external light conditions. However, the absolute control of light conditions is not possible since each camera uses a different LED; therefore, the color temperature and intensity may change, as well as the LED location on the camera's enclosure, which makes the sample enclosure device-dependent.

pH is defined as the decimal logarithm of the reciprocal of the hydrogen ion activity in a solution (McCarty and Vitz 2006). Paper-based color strips change color depending on the substances used to create the reactive. Some of the most common compounds used to create pH indicators are: phenolphthalein ($C_{20}H_{14}O_4$), which turns colorless in acidic solutions but pink in basic solutions; methyl red ($C_{15}H_{15}N_3O_2$), which is red for a pH below 4.4 and yellow for a pH over 6.2; and thymol blue ($C_{27}H_{30}O_5S$), which turns red below pH 1.2, yellow for pH between 2.8 and 8 and blue for pH above 9.6. These substances are combined to react to a specific range of pH. Universal paper strips are mixtures of several pH indicators that extend the pH range over which they operate. The molecular geometry of these chemical compounds changes according to the pH. This is due to modifications in the number of protons, which alter wavelength absorption causing color change.

Color evaluation has a wide field of applications. It can be used in drug dose control (Wickström *et al.* 2018), glucose detection (Dominguez *et al.* 2017), water quality control (Sicard *et al.*, 2015), gas detection (Escobedo *et al.* 2017) or pH determination (Loh *et al.* 2011). The most common way to digitize the color for further evaluation consists of using a photographic camera (Loh *et al.*, 2011; Oncescu *et al.* 2013; Kim *et al.* 2017). An alternative method for electronic color measurements is based on colorimeters. Colorimeters are tristimulus devices that use red, green and blue filters to emulate the response of the human eye to light and color. Recently, low-cost colorimeters integrated into circuit devices have become commercially available with good performance. Low-cost colorimeters can therefore be designed for several applications (Dominguez *et al.* 2017; Escobedo *et al.* 2017; Wickström *et al.* 2018). Unlike the problems associated with smartphone cameras, color measurement by colorimeter has the advantage of repeatability as a result of controlled light conditions.

Some references regarding pH monitoring systems based on NFC technology can be found in the literature. A low-power tag based on the SL13A chip from AMS with potentiometric input, to be used with pH and ion-selective electrodes, is presented in Kassal, Steinberg and Steinberg (2013). A computer-based reader controls the sampling acquisition. A recent study (Rahimi *et al.* 2017) shows a medical application for wound care by measuring the color of a pH reactive compound located on a bandage. The color is measured using a photodiode and associated information is stored on an MLX90129 NFC chip with data logger function from Melexis. These methods use a battery on the tag to sense the data and save it into the memory.

This chapter presents the third case study proposed in this book. It is a low-cost NFC-based system that combines an NFC IC for communication and a colorimeter IC with a microcontroller. The tag is completely passive and obtains its energy from the RF interrogating signal that comes from the reader using the energy-harvesting capabilities of the NFC IC. A calibration method is proposed and used to convert the measured color of the pH strips into pH levels. A custom readout application to operate in the smartphone, used as NFC reader, is also presented.

4.2. System description

4.2.1. *System overview*

An overview of the system is shown in Figure 4.1. The tag (or pH sensor) consists of an electronic circuit made up of a color light-to-digital TCS34725 converter from TAOS (TAOS 2012), a white 4150 K LED to illuminate the sample, a low-power Atmel ATtiny85 microcontroller and an NFC chip from ST Microelectronics (M24LR04E). The tag is read by means of a smartphone that can simultaneously upload the readings to the cloud in order to be stored.

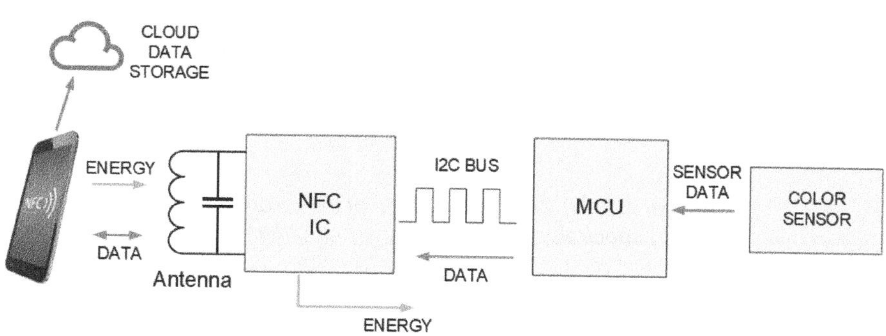

Figure 4.1. *Block diagram of the proposed NFC-based pH sensor*

The TCS34725 has RGB and clear light sensing elements (see Figure 4.2) and its light-to-digital converter contains a 3 × 4 photodiode array. This array comprises red-filtered, green-filtered, blue-filtered and clear (unfiltered) photodiodes. These photodiodes are coated with an IR-blocking filter that minimizes the IR spectral component of the incoming light, making it possible to obtain accurate color measurements. Spectral response curves of the four channels of the sensor are shown in Figure 4.2(b) (TAOS 2012). The four integrating ADCs simultaneously convert the amplified currents from the photodiode to a 16-bit digital value. Data is sent to the IC of the NFC via an I^2C bus, while the NFC IC is connected to the antenna consisting of a printed square spiral.

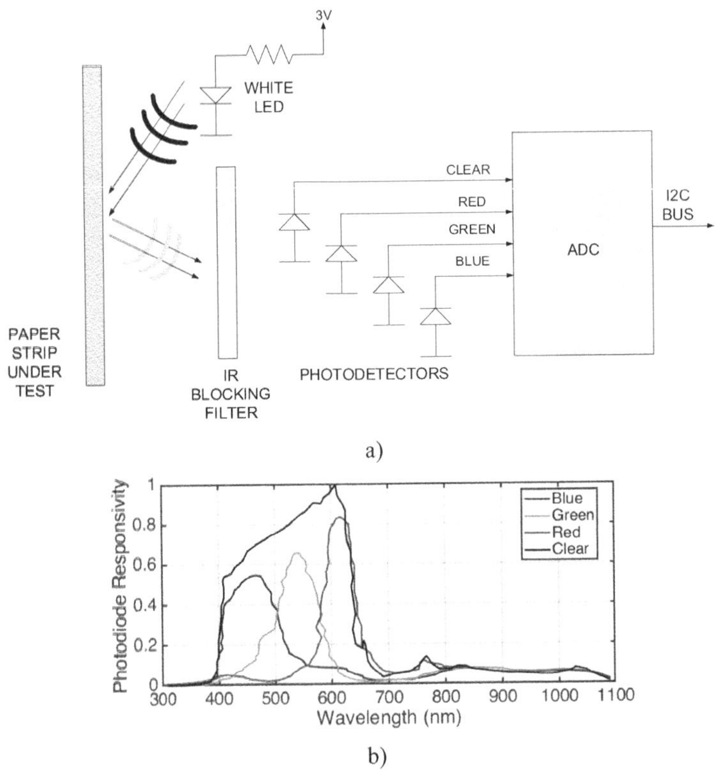

Figure 4.2. *(a) Working principle of the colorimeter; (b) spectral responsivity of the four channels*

The current consumption of each component is shown in Table 4.1. The NFC IC has four sink current configurable ranges. In this case, the mode with the highest sink current (up to 5 mA) was chosen to support the current needed for the tag operation. The Atmel 8-bit AVR ATtiny85 microcontroller can be set to different clock speeds and operates down to 1.8V. Because of the power limitations presented by this system, the microcontroller was configured to work at 1 MHz, where its consumption is around 300 μA at 3.3 V. The total current consumption of the tag is about 3 mA, which is less than the 5 mA that the NFC IC can harvest from the RF. A significant part of the overall consumption is due to the white LED used to illuminate the sample.

Component	**Current consumption (μA)**
M24	400
TCS34725	235
LED	2000
ATtiny85	300
Total	**2935**

Table 4.1. *Current consumption*

A photograph of the prototype is shown in Figure 4.3. The antenna, the NFC IC and the microcontroller are located on the top side of the PCB, while the color sensor and the LED are connected on the bottom side. The connector that interconnects the two circuit boards is also used to download the firmware in the microcontroller. A plastic enclosure was designed with a 3D printer to cover the tag (see Figure 4.3(c)).

a) b) c)

Figure 4.3. *Photograph of the prototype: (a) top view; (b) bottom view; and (c) 3D printed enclosure*

4.3. Signal processing

A mathematical model is needed to transform the raw color measurement in RGB provided by the sensor into pH. The first step is to define a color model in order to digitalize a color to represent or analyze it. RGB is the most common among the color models that exist. This additive color model is widely used for displays. CMYK is another color model that is widely used for color printing. Since color models are abstract mathematical models, it is possible to perform conversions from one model to another.

Figure 4.4(a) shows pH as a function of the color of the paper strip. The concentration of pH produces a change in color. One way to obtain the pH from the raw RGB measurement is to use a table-based method and an interpolation function. However, a simple model using HSV (Hue-Saturation-Value) color space is also available (Capel-Cuevas *et al.* 2012; Lopez-Ruiz *et al.* 2014). This model is useful not only because it is more intuitive than raw RGB values, but also because the conversions to and from RGB are extremely quick to compute and can be run in real time on the microcontroller. HSV is an alternative representation to RGB based on how the human eye works. Instead of measuring the portion of each primary color, the results are expressed in terms of Hue (representing pure color and expressed in degrees from 0° to 360°), Saturation (percentage of white color) and Value (brightness, expressed as a percentage), as can be observed in Figure 4.5. When color changes are to be measured, the HSV model will be the most appropriate and should be considered.

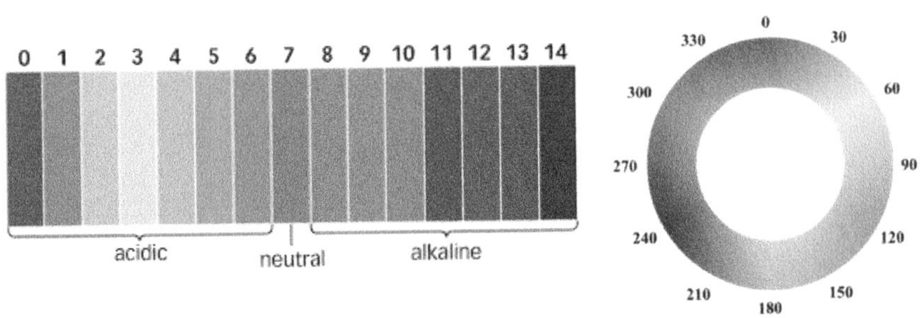

a) b)

Figure 4.4. *(a) pH as a function of the color of the paper strip; (b) hue. Primary colors are at 0° (red), 120° (green) and 240° (blue)*

From Figure 4.4(b), a relationship between hue (angle) and pH is expected. The transformation given in Pierre *et al.* (2015) to transform RGB into hue using equations [4.1] and [4.2] is used:

$$H = \begin{cases} \theta & \rightarrow if\ B \leq G \\ 360 - \theta & \rightarrow otherwise \end{cases} \qquad [4.1]$$

where θ (in degrees) represents the hue angle and R, G and B are the values of red, green and blue, respectively, in the 0–255 range.

$$\theta = arcos\left(\frac{\frac{1}{2}((R-G)+(R-B))}{\sqrt{(R-G)^2+(R-B)(G-B)}}\right) \qquad [4.2]$$

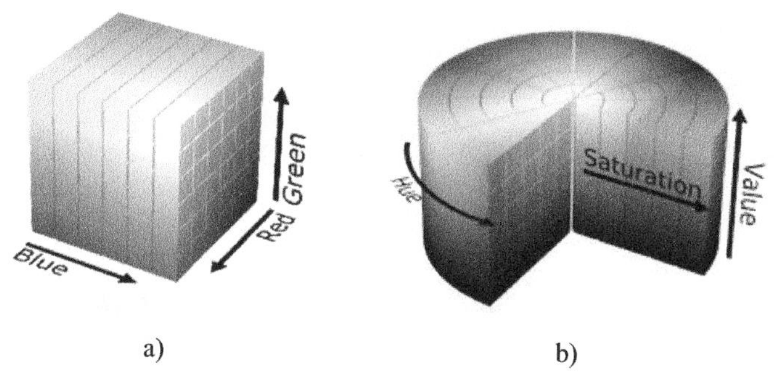

Figure 4.5. *(a) RGB color space; (b) HSV color space*

4.4. Experimental results

The hue obtained from measuring the color of pH reactive strips with the NFC colorimeter and the pH obtained using a calibrated pH probe (Hanna Checker HI 98103) are compared in order to test the system. The pH probe was previously calibrated using two reference solutions of pH 4 and 7. Three paper strips with different pH ranges were tested. Figure 4.6 shows the color charts of these pH paper strips. Figure 4.6(a) shows the color palette of the full range strips (pH from 1 to 14), while Figure 4.6(b) and (c) correspond to the pH ranges 3 to 6 and 5 to 9, respectively.

Several values of pH are measured for each strip (see Figure 4.7). This figure shows linear behavior in the measured ranges. Thanks to the linear behavior of the hue of the strips, it is possible to create a linear regression model (the red line in Figure 4.7) to characterize each type of strip. Based on this linear model, the system can be calibrated by measuring two different pH values.

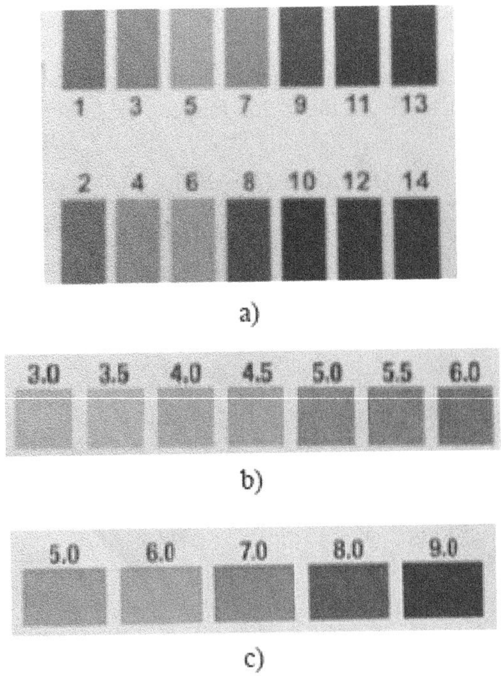

Figure 4.6. *Color charts for the three paper strips tested: (a) chart for pH scale 1 to 14; (b) chart for pH scale 3 to 6; and (c) chart for pH scale 5 to 9*

Because most current methods for determining the color of reactive strips are based on the image processing of a photograph of the strips (Vuong *et al.* 2010; Loh *et al.* 2011; Kim *et al.* 2017), both the sensor color reading and the image processed by two different cameras are compared. The main disadvantages of image processing systems are the variability of the brightness of the samples and the color correction applied to the pictures by each vendor. In order to compare the two methods, 30 measurements of four different strips are done. Figure 4.8 shows the hue calculated by placing and removing the strips from the 3D-printed enclosure (the black line with diamonds) 30 times. The red and blue lines show the results of 30 pictures taken for each case at several distances (between 20 and 50 cm) and with slight angle changes. The images are taken by two different cameras in direct and indirect light conditions for each one (continuous and dashed lines, respectively).

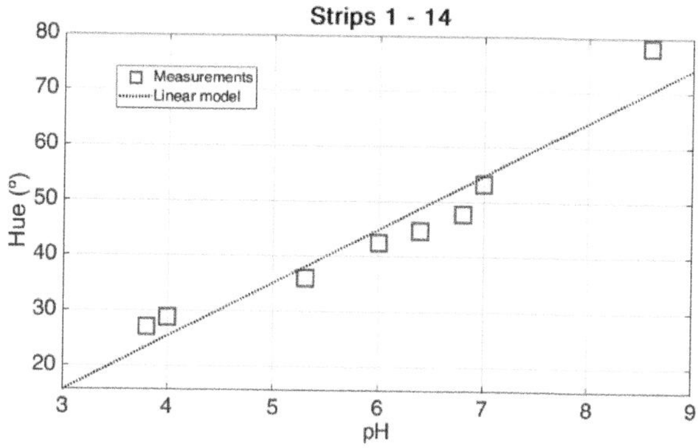

a)

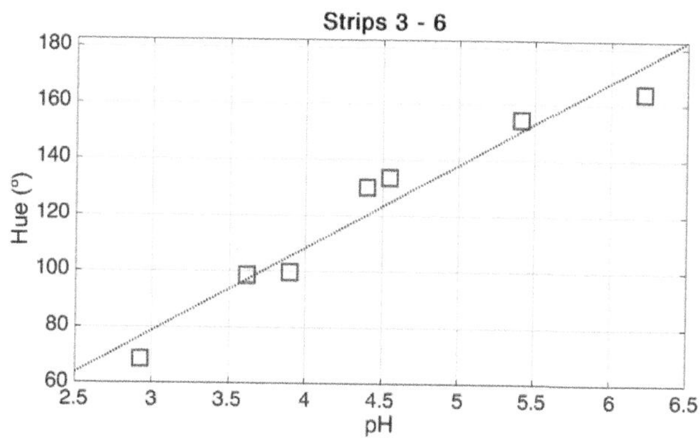

b)

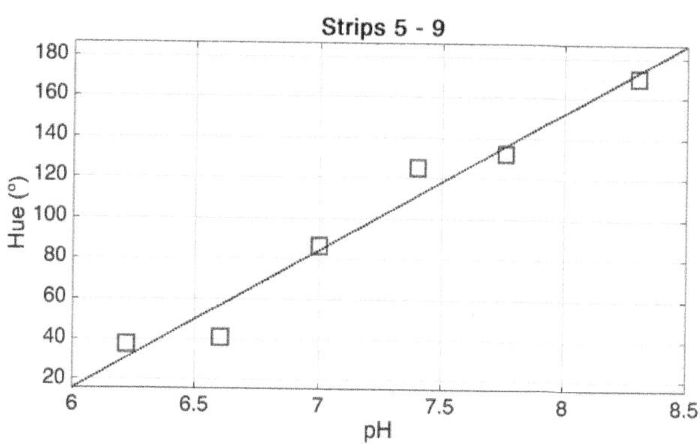

c)

Figure 4.7. *Hue measurements of several known pH values. (a) Strips pH 1 to 14; (b) strips pH 3 to 6; and (c) strips pH 5 to 9*

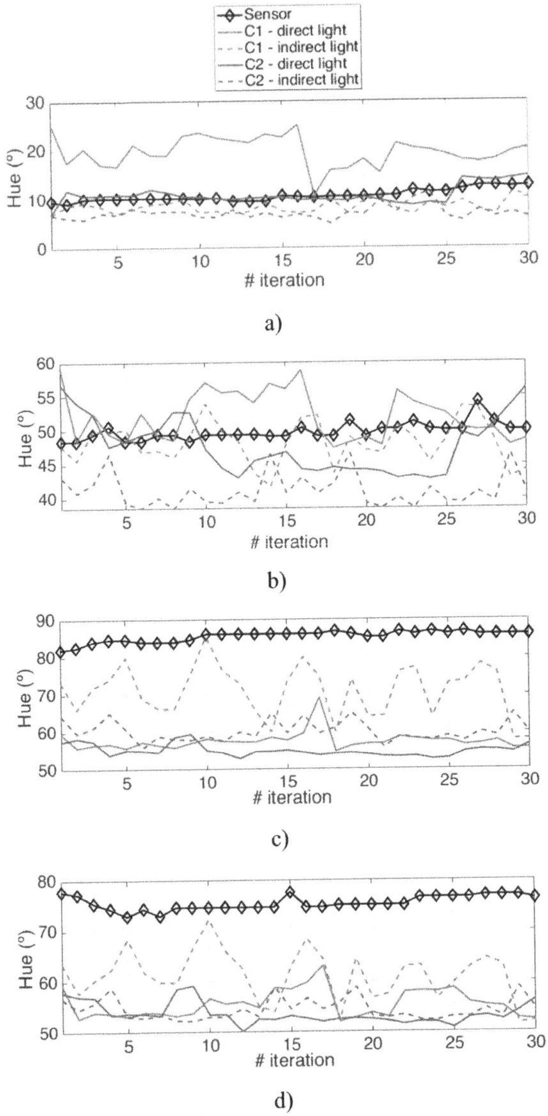

Figure 4.8. *Comparison of hue angle for 30 iterations of four different cases. (a) Strips from 1 to 14: pH 4; (b) strips from 1 to 14: pH 7; (c) strips from 3 to 6: pH 4; and (d) strips from 5 to 9: pH 7*

The pictures are taken at the maximum resolution for each camera (5488×4112 pixels for camera 1 and 5120×3840 pixels for camera 2). The images are analyzed using MATLAB. The exact reactive part of the strips is selected manually. The average RGB is measured and converted to HSV using equations [4.1] and [4.2]. This is, therefore, an identical procedure to that used by image processing systems.

It can be seen how the iterations measured with this system behave linearly, since the lighting conditions and the distance between the sensor and the samples to be measured are always the same. On the contrary, for color analysis based on photography, these parameters are much more unstable.

Table 4.2 shows the standard deviation for each sample and each sensing method. The effects of different lighting conditions and the deviation from accurate focus are quantified. As the mean deviation values show, the results from camera 2 are better than those from camera 1. However, the best results are obtained with the sensor proposed in this chapter.

	Sample 1	**Sample 2**	**Sample 3**	**Sample 4**	**Mean deviation**
Sensor	1.01	1.16	1.25	1.27	**1.17**
C1 direct light	3.19	3.62	2.47	2.74	**3.00**
C1 indirect light	1.27	3.16	7.03	5.08	**4.14**
C2 direct light	1.82	4.23	1.76	2.24	**2.51**
C2 indirect light	1.34	2.49	2.73	2.14	**2.18**

Table 4.2. *Hue standard deviation (degrees)*

To graphically represent the deviation observed when the same sample is measured several times, Figure 4.9 shows the histograms of the values obtained. In all cases, the sensor proposed in this chapter obtained the best results.

Of course, light plays the most important role when measuring color but other factors, including camera optics, image focusing and processing software can also affect the reading. Moreover, systems based on image processing need complex software to detect the part of the image that will be analyzed. In the system proposed in this chapter, the conditions are always the same and the smartphone used to obtain the pH has no effect on the color reading; therefore, the best results are obtained. It is therefore device-independent, does not depend on conditions in the surrounding environment, and is simpler than other methods for determining the color of the reactive strips.

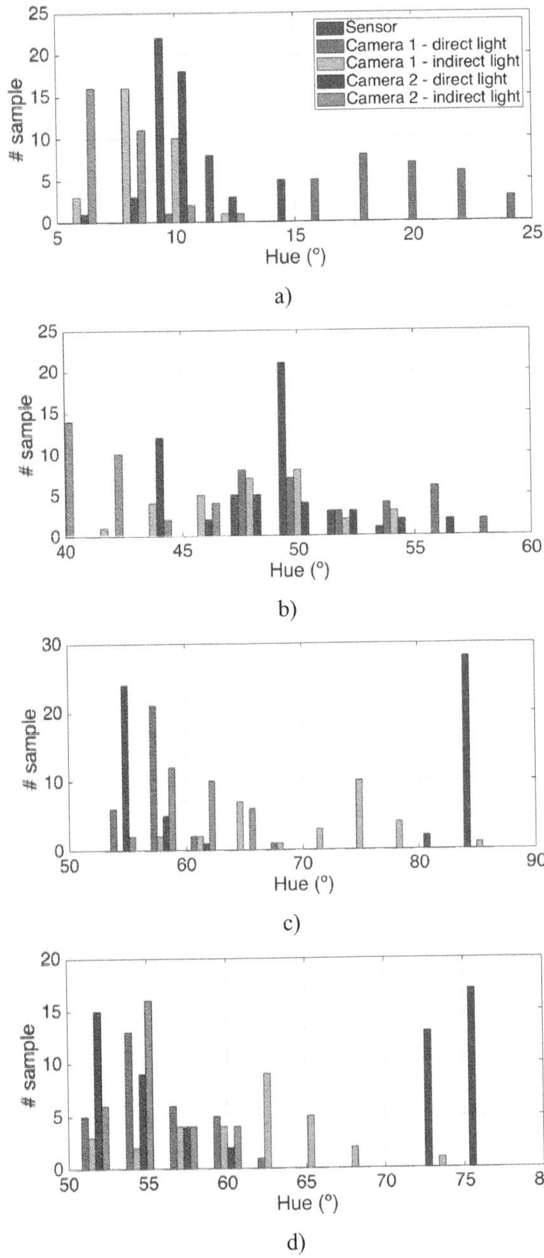

Figure 4.9. *Histograms of measured values for four different cases. (a) Strips from 1 to 14: pH 4; (b) strips from 1 to 14: pH 7; (c) strips from 3 to 6: pH 4; and (d) strips from 5 to 9: pH 7*

4.5. App and cloud storage

An Android application (app) to display the results has been designed. This application reads the HSV values using the NFC protocol. From the stored linear model of section 4.4, it returns the pH value. The app also includes an option to calibrate strips from different manufacturers. For this purpose, measurements of two known pH values must be taken.

The pH data is uploaded to a cloud database in order to store all historical measurements, and for additional analysis and statistics. Figure 4.10(a) shows an image of the application's main screen, while Figure 4.10(b) shows the calibration screen that indicates the two known pH values used to calculate the linear function that are used to estimate the pH of the measurements.

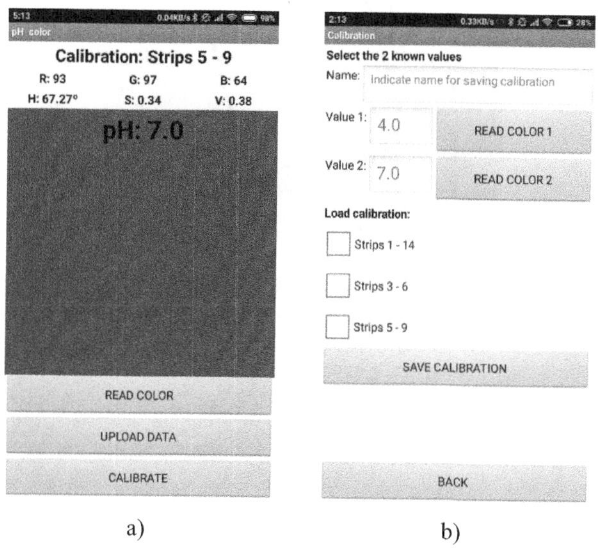

Figure 4.10. *Designed Android app: (a) main screen and (b) calibration screen*

Figure 4.11 shows the device placed inside its 3D-printed enclosure, with a pH strip in the measurement slot. It also shows the smartphone used to power up the device, and retrieve and display data using the designed application. The theoretical and experimental read range obtained for the tag prototype is up to 18 mm depending on the smartphone used.

Figure 4.12 shows a graph of the results uploaded to the ThingSpeak IoT cloud database, which shows several measurements of the pH performed in a swimming pool. This is an example of a weekly water analysis performed by the pool's maintenance staff. Historical measurements can thus be accessed from any device connected to the Internet. In this way, it can be guaranteed that the pH is kept within an acceptable range for the safety of the pool, by monitoring it periodically.

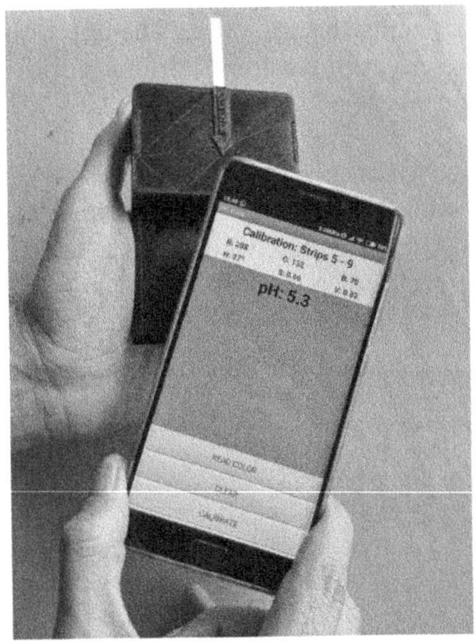

Figure 4.11. *Reading the color of a strip and calculation of the pH with the Android application*

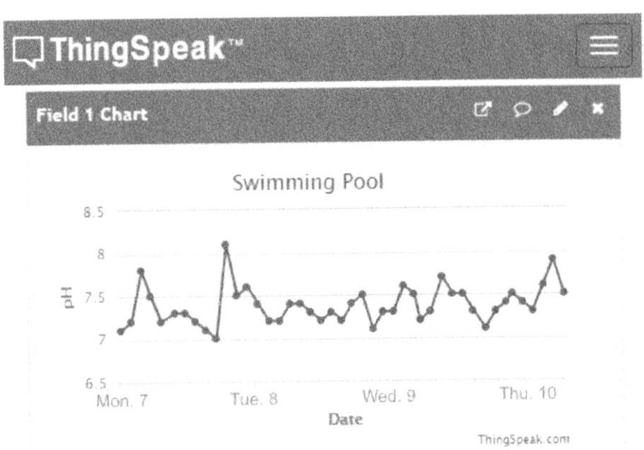

Figure 4.12. *Cloud database showing the pH measurements of a swimming pool taken over several months*

5

Case Study 4: Fruit Ripeness Sensor

5.1. Motivation

The color of food is the first sensation perceived by the consumer that determines its appearance, influencing their subsequent choice. Food color depends on various factors, including the chemical, biochemical, microbial and physical changes that occur during growth, maturation, postharvest handling and processing (Pathare *et al.* 2013). Therefore, food color can be used as a quality indicator. Color is one of the most widely measured product quality attributes in postharvest handling and in food processing research and industry.

The interest of consumers in organic food, to know its origin and traceability, has grown in recent years. In this context, RFID technology can play an important role in the coming years. Today, the long market chains that affect many types of fruits and vegetables make it difficult to establish the use-by date or recommended consumption date of these products. Consequently, surface color can be a good indicator of the freshness and flavor quality of fresh products. For example, the ripening process of tomatoes is clearly characterized by the evolution of the color on the tomato's surface – after harvesting, the fruit continues to ripen and its color turns from green to red. Because transport time must be taken into account, tomatoes that are already ripe (red) must be sold to local markets, while green ones can be shipped to customers located at much greater distances (Lee *et al.* 2011). Color has been used to evaluate fruit maturity for tomatoes (Syahrir *et al.* 2009), bananas (Mustafa *et al.* 2008), apples (Rennick *et al.* 1999), pears (Zhao *et al.* 2009), oranges (Recce *et al.* 1996) and dates (Lee *et al.* 2008). Some colors are associated with food products. For instance, yellow is associated with ripe bananas in good conditions, and good tomatoes are associated with red rather than orange.

For a color version of all the figures in this chapter, see: www.iste.co.uk/girbau/sensors.zip.

Food color measurements can be performed using a wide variety of instruments (Pathare *et al.* 2013) such as colorimeters (Mendoza *et al.* 2006) and spectrophotometers (HunterLab 2008; Vandekinderen *et al.* 2008), as well as by applying computer vision techniques using digital cameras (Scanlon *et al.* 1994; Recce *et al.* 1996; Rennick *et al.* 1999; Lee *et al.* 2008; Mustafa *et al.* 2008; Syahrir *et al.* 2009; Zhao *et al.* 2009). However, sometimes these instruments are expensive, or the measurement setups are not suitable for performing real-time measurements by a consumer in a supermarket or at home.

Colorimeters and spectrophotometers are devices that can be used to measure the color of a test sample. A spectrophotometer is a high-precision instrument, and it is suitable for performing more complex color colorimetric analysis because it can measure the spectral reflectance at each wavelength. However, spectrophotometers are more expensive than colorimeters. Therefore, a colorimeter is the best choice for quality inspection.

In this chapter, a low-cost battery-less NFC tag is presented for food color measurement and classification. It is based on the measured color and the use of a smartphone as a reader. It is possible to read the information from the colorimeter integrated in the label, by means of a specific application installed in an NFC-equipped smartphone. This low-cost device can be used to classify different fresh products in the supermarket or at home.

It must be highlighted that this application has been selected, since it allows us to demonstrate the combination of NFC technology to power and read information from a sensor with the use of advanced processing algorithms based on machine learning of low computational complexity that can be implemented in a mobile environment.

5.2. System overview

A system overview is shown in Figure 5.1. The food being tested is placed over the aperture of a colorimeter integrated in the NFC tag. Its color is a function of the temperature, humidity and days since it was harvested. When the user taps their smartphone, the tag's NFC IC harvests the energy from the RF interrogating signal, which is used to power the circuitry and then returns the color measurement. The application in the smartphone reads the message, classifies the quality of the food and gives additional information about the origin or some other useful information. This information, as well as the parameters needed for food classification, can be automatically downloaded from the cloud.

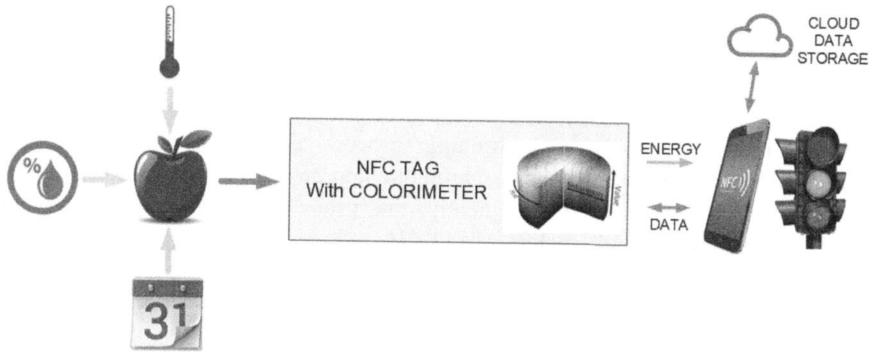

Figure 5.1. *System overview*

The tag is identical to the one presented in the previous chapter, composed by an M24LR04E NFC IC, an ATtiny85 as a microcontroller, and the color sensor TCS34725, with a total consumption of almost 3 mA (see Table 4.1). In this work, a 3D printed housing for the tag with an opening for the color sensor has been designed (Figure 5.2).

Figure 5.2. *Tag inside the 3D printed housing*

5.3. Experimental results

In order to test the system, different typical fruits (bananas, and red and golden apples) are measured with the NFC colorimeter. The objective is to classify the pieces of fruit according to their quality based on the number of days outside the refrigerator using the color information. To this end, several samples stored at room temperature and others stored in the refrigerator for several days were tested. Any fruit is considered to belong to the good class if the days at room temperature outside the refrigerator are equal to or less than six. The aim is to find a simple but accurate method that, after a training process performed by the product manufacturer, can be installed in a smartphone app and the consumer can run it to evaluate the fruit's quality. Consequently, this processing method should not be based on sophisticated tools such as MATLAB.

As already introduced in Chapter 5, the color of an object can be described by several color coordinate systems (called color spaces). Figure 5.3 shows the representation of the RGB (red, green and blue), HSV (hue, saturation, value) and CIELab color spaces. The first decision is to choose the color space. The raw RGB information given by a colorimeter is rarely used in the literature for food classification. HSV (Pierre *et al.* 2015) and Lab (Pierre *et al.* 2015) are alternative representations of the RGB color model that more closely align with the way human vision perceives colors. HSV and Lab color spaces are some of the most used in the literature for this kind of test (Chen *et al.* 2007; Ke *et al.* 2013; Trambadia and Mayatra 2017). The more frequently used however is the CIELab color space, due to its uniform color distribution and because its color perception is closer to that of the human eye. The color in the CIELab color space is determined by three values: L* is the lightness, and a* and b* are the green-red and blue-yellow color components, respectively. Humans identify a colored object through its chromaticity and brightness. The chromaticity can be further divided into two parts: hue and saturation.

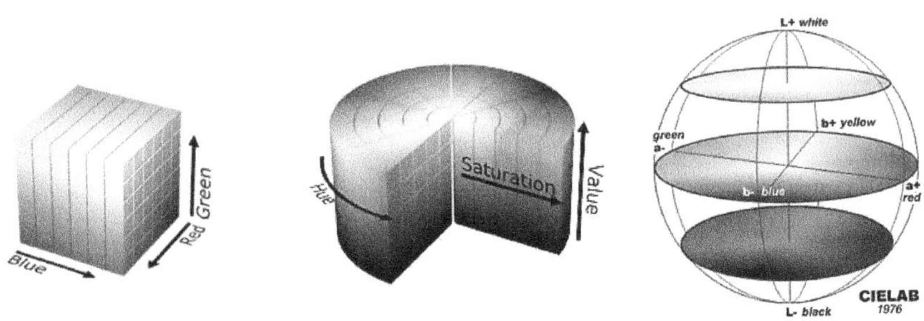

a) b) c)

Figure 5.3. *(a) RGB color space; (b) HSV color space; and (c) CIELab color space*

The a/b ratio was used as a color index in apples, tomatoes, and citrus and carambola fruit (Pathare *et al.* 2013). A high correlation between peel color related to maturity and Hunter's a/b ratio for mangos has been found in the work of

MEDLICOTT *et al.* (1992). However, this ratio is a function of the hue, which is the angle that can be computed from the *atan*(a/b) relationship. Therefore, this model will be used in this chapter.

Figure 5.4 shows the histograms of HSV measurements of a golden apple as a function of the days and the ripening conditions (in the refrigerator and at room temperature, respectively).

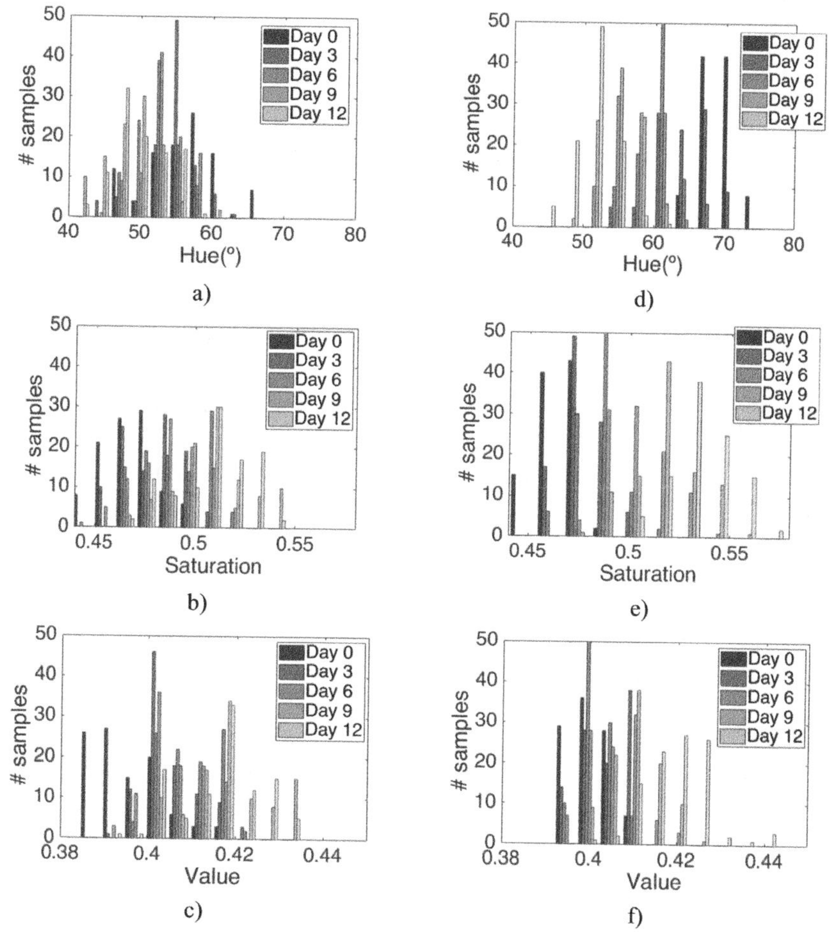

Figure 5.4. *Histograms for a golden apple for different days: histogram of the hue (a), saturation (b) and value (c) parameters in the refrigerator, and histogram of the hue (d), saturation (e) and value (f) parameters at room temperature*

Three kinds of fruit (bananas, red apples and golden apples) are considered. The measurements are taken with the colorimeter of the NFC tag. The Cumulative

Distribution Function (CDF) of the hue for the golden apple in the refrigerator and at room temperature as a function of the number of days is shown in Figure 5.7. It can be seen that the color variation is slight in the case of the fruit preserved in the refrigerator and even faster when stored at room temperature. Therefore, the ripeness grade is also a function of environment parameters. The histograms of HSV measurements and CDF of the hue parameter for the banana (Figures 5.5 and 5.8) and red apple (Figures 5.6 and 5.9) are shown. Similar behavior is noted for both the banana and red apple.

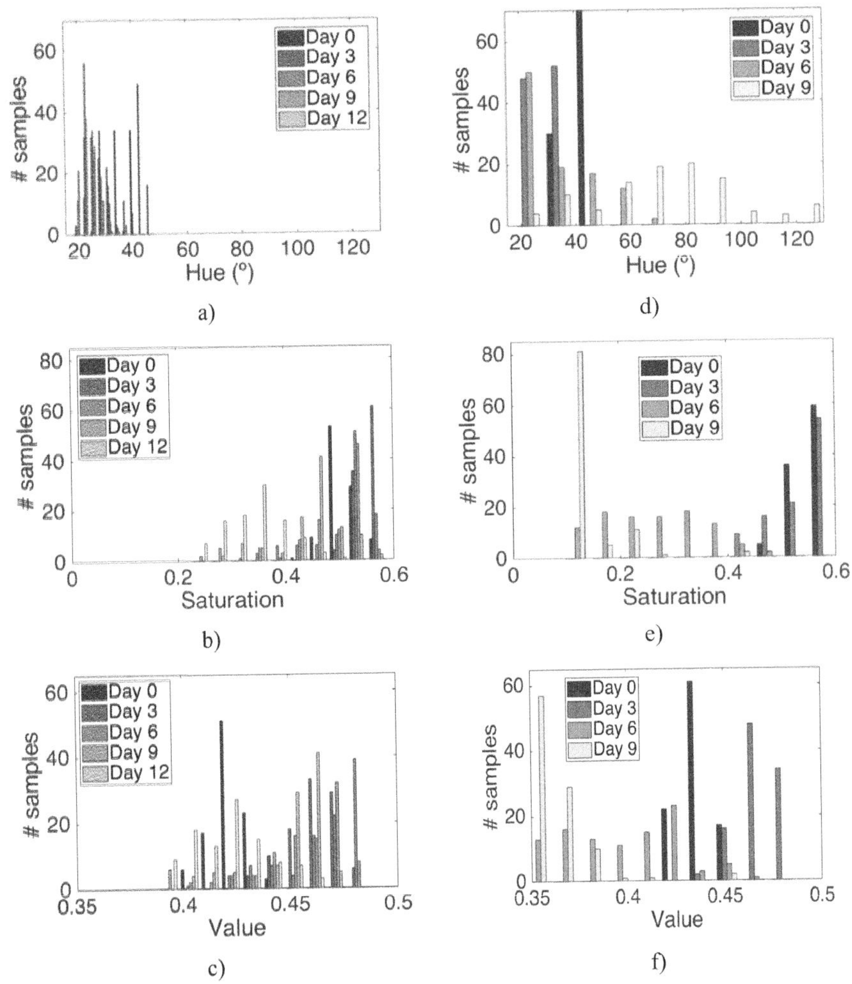

Figure 5.5. *Histograms for a banana for different days: histogram of the hue (a), saturation (b) and value (c) parameters in the refrigerator, and histogram of the hue (d), saturation (e) and value (f) parameters at room temperature*

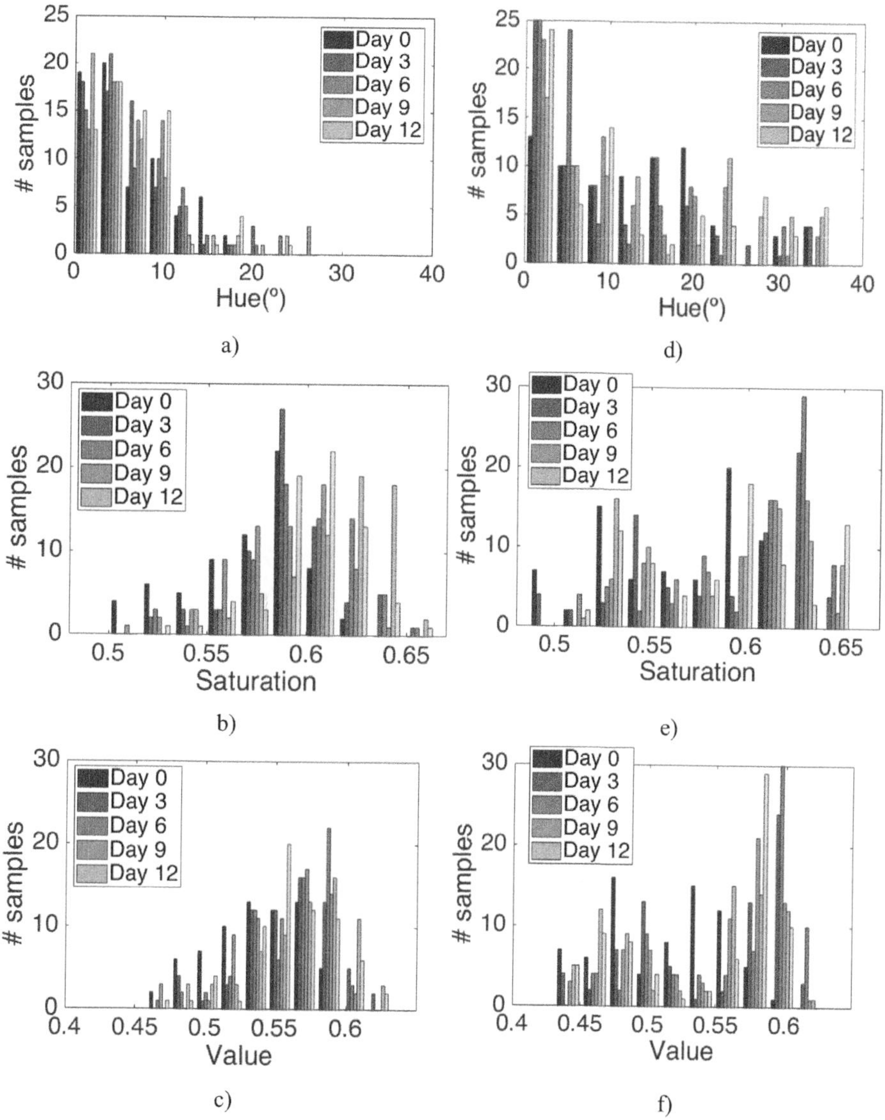

Figure 5.6. *Histograms for a red apple for different days: histogram of the hue (a), saturation (b) and value (c) parameters in the refrigerator, and histogram of the hue (d), saturation (e) and value (f) parameters at room temperature*

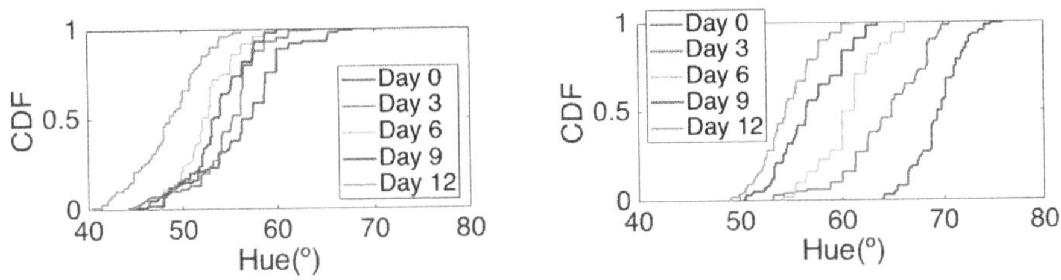

a) b)

Figure 5.7. *Cumulative Distribution Function (CDF) of the hue parameter for the golden apple in the refrigerator (a) and at room temperature (b) as a function of the number of days*

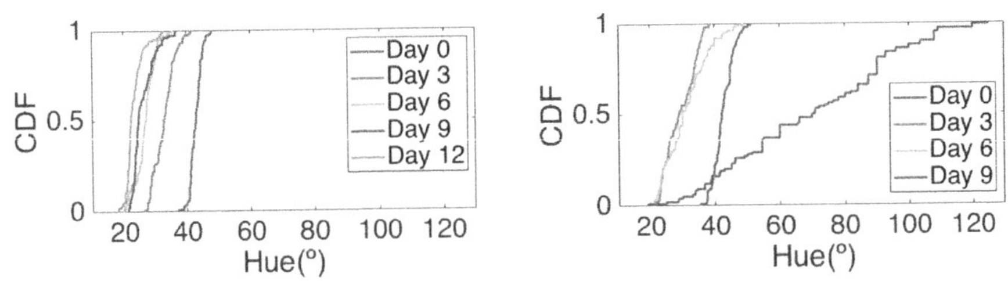

a) b)

Figure 5.8. *CDF of the hue parameter for the banana in the refrigerator (a) and at room temperature (b) as a function of the number of days*

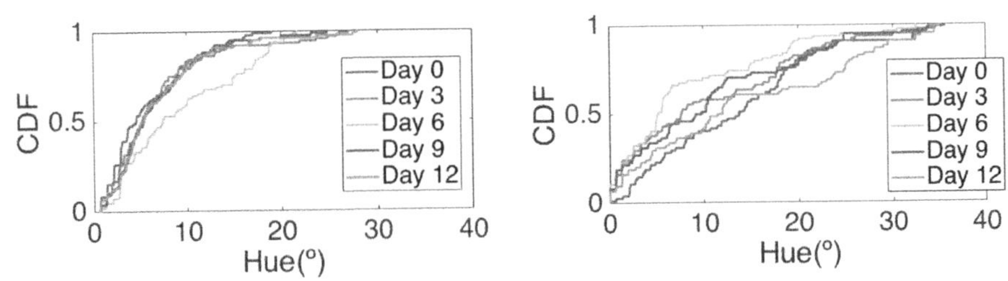

a) b)

Figure 5.9. *CDF of the hue parameter for the red apple in the refrigerator (a) and at room temperature (b) as a function of the number of days*

In order to select the most suitable values to analyze the ripeness of the fruit, HSV and CIELab values are studied. Table 5.1 shows the average of 100 samples for each fruit stored at room temperature, comparing the measurements of the first day (day 0) with those obtained after a few days (day 15 for apples, and day 9 for bananas). The Δ% row shows the percentage increment of each parameter, considering its range. We observe that the best combination is obtained from the values H and S. The hue changes 8.7% for red apples, 5.4% for golden apples and 1.6% for bananas; whereas the saturation changes 42% for red apples, 12% for golden apples and 2% for bananas. However, the variations are smaller using CIELab color space, with differences ranging between 0.1% and 2% for L, 1.2% and 4.3% for parameter a* and 1.4% and 8.2% for parameter b*.

Fruit	Days	H	S	V	L*	a*	b*
Golden apple	Day 0	59.4	0.48	0.40	41.1	-7.3	26.6
	Day 15	40.0	0.60	0.46	40.9	3.6	30.2
	Δ %	5.4	12.00	6.00	0.1	-4.3	1.4
Banana	Day 0	17.4	0.58	0.51	37.9	20.6	22.1
	Day 15	23.2	0.60	0.51	39.4	17.6	25.8
	Δ %	1.6	2.00	0.00	1.5	1.2	1.4
Red apple	Day 0	41.2	0.55	0.44	40.0	2.0	27.1
	Day 9	72.5	0.13	0.36	38.0	-2.4	6.0
	Δ %	8.7	42.00	8.00	2.0	-1.7	8.2

Table 5.1. *Average readings of HSV and L*, a* and b* for 100 samples. Comparative change of each parameter over time. H = hue, S = saturation, V = value, L* = lightness, a* and b* = chromaticity coordinates. a* and b* are color directions: +a* is the red axis, -a* is the green axis. +b* is the yellow axis and -b* is the blue axis*

Extracting and selecting features helps to improve the machine learning algorithm by focusing on the data that are most likely to produce accurate results. The challenge is to find the minimum number of features that will capture the essential patterns in the data. From the histograms presented above, it can be concluded that examining the hue and saturation are good options. The variation over time is higher compared to the value parameter. Another possibility is to apply PCA (Principal Component Analysis) to find the principal components in order to reduce the number of features. However, PCA is not considered to simplify the implementation in a mobile environment because it can increase the computational complexity.

The next step is to select a machine learning algorithm. No single machine learning algorithm works for every problem; therefore, the best algorithm is found by exploring the data sets for different algorithms. The comparisons were performed using the statistics and machine learning toolbox (MATLAB R2017a software). The training is performed using a computer with MATLAB; this step ideally should be performed by the provider or a laboratory. After the training is finished, the parameters are sent to the cloud server, from where the consumer can download the parameters that are necessary to predict the class. The simplicity of the algorithm and the minimum tuning play an important role as the prediction step will be made on a smartphone. A second option is to execute a program in the cloud server that returns the results as a function of the color measurement. However, the computation on the smartphone is the preferred option. Two classes were considered, assuming that the fruit is outside the refrigerator: good (days <6) and bad state (days> =6). Other intermediate classes can be defined if necessary. Half of the measurements in the data set were used for training, and the other half were used for testing the prediction. In this chapter, the following algorithms are considered and compared: linear SVM (support vector machine) (Andrew 2001), linear discriminate analysis (LDA) (Mika *et al.* 1999), Naive Bayes (Rish 2001), decision tree (Breiman *et al.* 2017) and nearest neighbor (NN) (Muja and Lowe, 2009). Neural networks are not included because they often require many tuning parameters.

Although SVM and LDA have the same boundary decision (a line), their assumptions are different. LDA assumes that data are distributed normally. SVM makes no assumptions about the data at all, which means that it is a very flexible method. LDA makes use of the entire data set to estimate covariance matrices, and therefore, it is somewhat prone to outliers. SVM is optimized over a data subset that includes those data points that lie on the separating margin. The data points used for optimization are called support vectors because they determine how the SVM discriminates between groups and thus supports the classification.

The naive Bayes classifier is designed to be used when predictors are independent of each other within each class; however, it seems to work well in practice even when the independence assumption is invalid. It is a classification technique based on the Bayes theorem with an assumption of independence among predictors. In simple terms, a naive Bayes classifier assumes that the presence of a particular feature in a class is unrelated to the presence of any other feature.

The decision tree classifier is a simple and widely used classification technique. It repetitively divides the working area (plot) into subparts by identifying lines. In general, decision trees are constructed via an algorithmic approach that identifies ways to split a data set based on different conditions. It is one of the most widely used and practical methods for supervised learning. Decision trees are a non-parametric supervised learning method used for both classification and regression tasks. The goal is to create a model that predicts the value of a target variable by learning simple decision rules inferred from the data features. Decision trees divide the feature space into axis-parallel rectangles or hyperplanes.

The nearest neighbor search locates the k-nearest neighbors or all neighbors within a specified distance from certain query data points, based on the specified distance metric. An object is classified by a majority vote of its neighbors, and the object is assigned to the most-common class among its k-nearest neighbors (k is a positive integer, typically small). If k = 1, then the object is simply assigned to the class of that single nearest neighbor. One of the simplest decision procedures that can be used for classification is the nearest neighbor rule. This classifies a sample based on the category of its nearest neighbor.

In order not to introduce any bias in the classification results, the data set is composed of the same number of measurements each day. Another fruit that is not included in the training data set is used for testing. Figures 5.10–5.12 show the decision regions found for each algorithm superposed on the scatter plot of the training measurements of the data set.

To explore the trade-off between different kinds of misclassification, a confusion matrix is used. Table 5.2 summarizes the confusion matrix calculated with different methods: linear SVM, naive Bayes, decision tree and nearest neighbor, showing the true positives (*TP*), false positive (*FP*), false negative (*FN*) and true negative (*TN*) percentages. In order to compare the methods, the accuracy is also included. The accuracy of the confusion matrix is computed using equation [5.1] and the results for each feature are illustrated in the last column in Table 5.2.

$$Accuracy(\%) = 100\frac{TP + TN}{TOTAL\ DATA} \qquad [5.1]$$

The highest accuracy is obtained for the banana and the worst for the red apple. In the case of the golden apple, the two features (hue and saturation) are clearly correlated because all the training points fall along a line. Therefore, a single feature (e.g. the hue or using PCA decomposition) can be used for classification. A smooth variation of the color is found as a function of the time. The decision limit found for all classifiers is a line perpendicular to this line. Accuracies of about 82%–84% are obtained depending on the classifier algorithm considered. In the case of the banana, the color remains nearly constant for the first days, and then begins to degrade fast from the 6th day, as can be seen in the histogram. The change in color occurs in both the hue and saturation features, and the relation is not as linear as it is in the case of the golden apple. However, the two classes are clearly separable and high accuracy is obtained for all classifiers. The last case is more complicated to classify because the color is not homogeneous. This fruit has regions of red color; however, other regions are close to yellow. To avoid this problem, only the red region is considered, filtering the points before training in the range of hue <50°. If a point falls outside this range, the application asks the user to repeat the measurement at another point in the red region. After outliers have been filtered, the color degradation is smooth, as in the case of the golden apple. However, due to the heterogeneous distribution of the color on the surface, the accuracy obtained for the different classifier algorithms is considerably lower than in the previous cases. Only LDA and the nearest neighbor with k = 5 obtain reasonable values of 73% and 71%, respectively. An improvement is found if a quadratic discriminant analysis (QDA) is used, achieving an accuracy of 80% because the boundary decision can be better described for a parabolic function.

Fruit	Classifier	TP %	FP %	FN %	TN %	Accuracy %
Golden apple	Naive Bayes	73.66	6.67	26.33	93.33	83.50
	Linear discriminant analysis	73.00	8.67	27.00	91.33	82.17
	Decision tree	76.33	12.33	23.67	87.67	82.00
	Nearest neighbor	81.33	13.33	18.67	86.67	84.00
	Nearest neighbor k = 5	80.00	13.67	20.00	86.33	83.17
	SVM	68.00	2.00	32.00	98.00	83.00
Banana	Naive Bayes	90.00	4.00	10.00	96.00	93.00
	Linear discriminant analysis	90.50	5.00	9.50	95.00	92.75
	Decision tree	84.50	3.50	15.50	96.50	90.50
	Nearest neighbor	87.00	13.00	13.00	87.00	87.00
	Nearest neighbor k = 5	87.00	3.00	13.00	97.00	92.00
	SVM	90.50	5.50	9.50	94.50	92.50
Red apple	Naive Bayes	73.33	46.67	26.67	53.33	63.33
	Linear discriminant analysis	86.67	40.00	13.33	60.00	73.33
	Decision tree	63.33	40.00	36.37	60.00	61.67
	Nearest neighbor	60.00	66.67	40.00	33.33	46.67
	Nearest neighbor k = 5	80.00	36.67	20.00	63.33	71.67
	SVM	96.67	90.00	3.3	10.00	53.33
	QDA	86.67	26.67	13.33	73.33	80.00

Table 5.2. *Confusion matrix for ripeness grading identification for different classifiers*

Machine learning applications can be deployed in production systems on desktops, in enterprise IT systems (either onsite or in the cloud) and in embedded systems. For prediction, the linear SVM, LDA or QDA give an analytical boundary decision function (a line or quadratic function) that only needs to know the polynomial coefficients to define the decision region. The evaluation of the algorithm in the other methods requires the training data from the server. In addition, it requires the implementation in JAVA, C, PHP or another language in the mobile environment or in the cloud. In order to avoid this problem, a table (or image) can be downloaded from the cloud service that is used to interpolate the results.

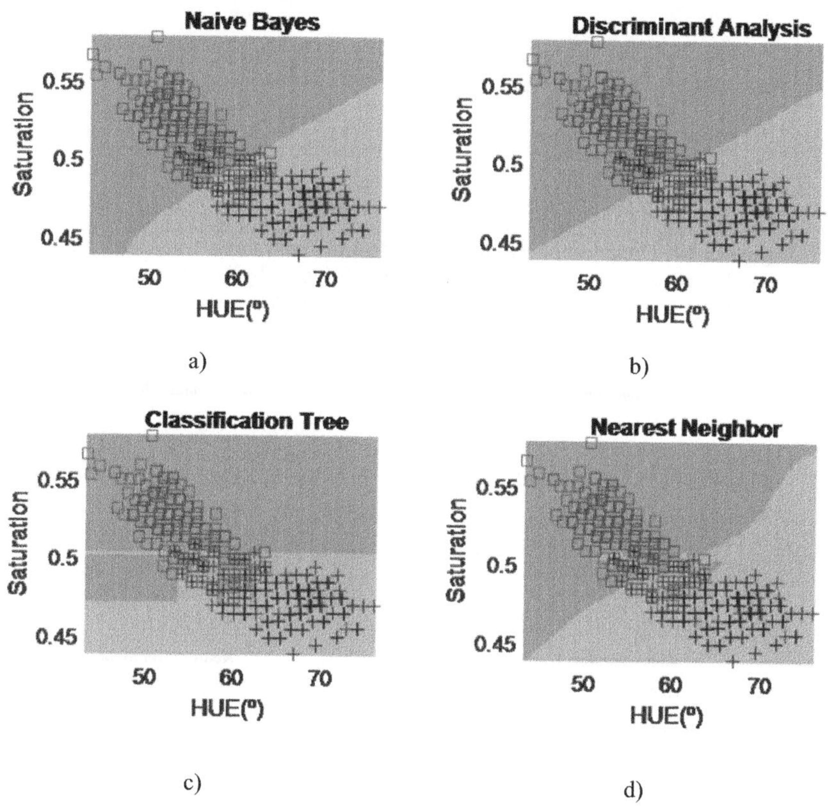

Figure 5.10. *Decision boundaries and scatter plot (class 1 squares and class 2, crosses) for the golden apple. (a) Naive Bayes; (b) linear discriminant analysis; (c) classification tree; and (d) nearest neighbor (k = 5)*

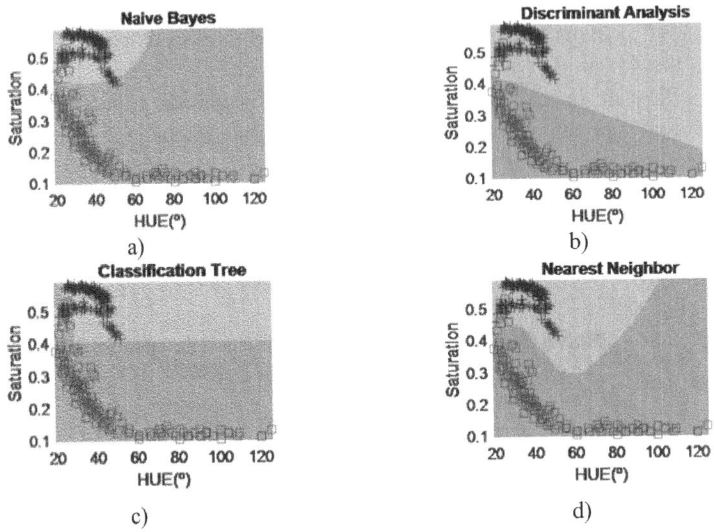

Figure 5.11. *Decision boundaries and scatter plot (class 1 squares and class 2, crosses) for the banana. (a) Naive Bayes; (b) linear discriminant analysis; (c) classification tree; and (d) nearest neighbor (k = 5)*

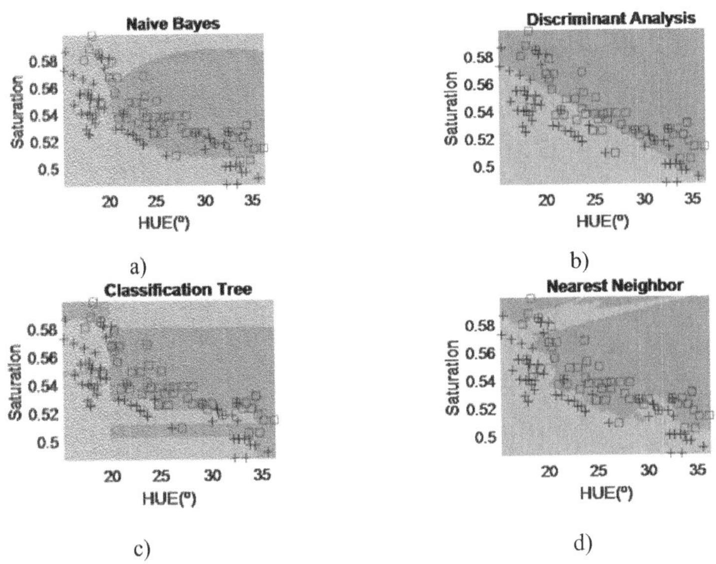

Figure 5.12. *Decision boundaries and scatter plot (class 1 squares and class 2, crosses) for the red apple. (a) Naive Bayes; (b) linear discriminant analysis; (c) classification tree; and (d) nearest neighbor (k = 5)*

5.4. Mobile app

An Android application has been developed to analyze the sample and show the result to the user. The operation flow is described in the flowchart of Figure 5.13.

When the application is launched, it connects to a server to obtain a list of available fruits, and downloads an image and information about each of them. Then, it shows the list of possible fruits to be analyzed (Figure 5.14(a)). The user selects the fruit and a message appears to place the smartphone over the tag (Figure 5.14(b)). When the tag is triggered, it performs the color reading, and sends the NDEF (NFC (near-field communication) data exchange format) message with the HSV (hue, saturation and value) and RGB (red, green and blue) values to the smartphone, which processes the message and shows the result on the screen (Figure 5.14(c)). The hue and saturation values are calculated to extrapolate the point using the image previously downloaded, which represents the decision boundaries of the training (Figure 5.14(d)). The application obtains the resolution and the minimum and maximum values of the hue and saturation for each image. This makes it possible to check the color of a specific point of the classification matrix and determine the class that the measured color belongs to.

This method is computationally cheap on the reader's side. Furthermore, the fact that the axis boundaries and the image resolution are retrieved from the server makes it possible to adapt this system to any kind of image, no matter which classification algorithm has been used or how many classes it contains, whereas the colors of each class are well differentiated.

Figure 5.15 shows a smartphone with the application in the screen after measuring the color of a red apple, which is placed on top of the 3D printed housing that contains the tag.

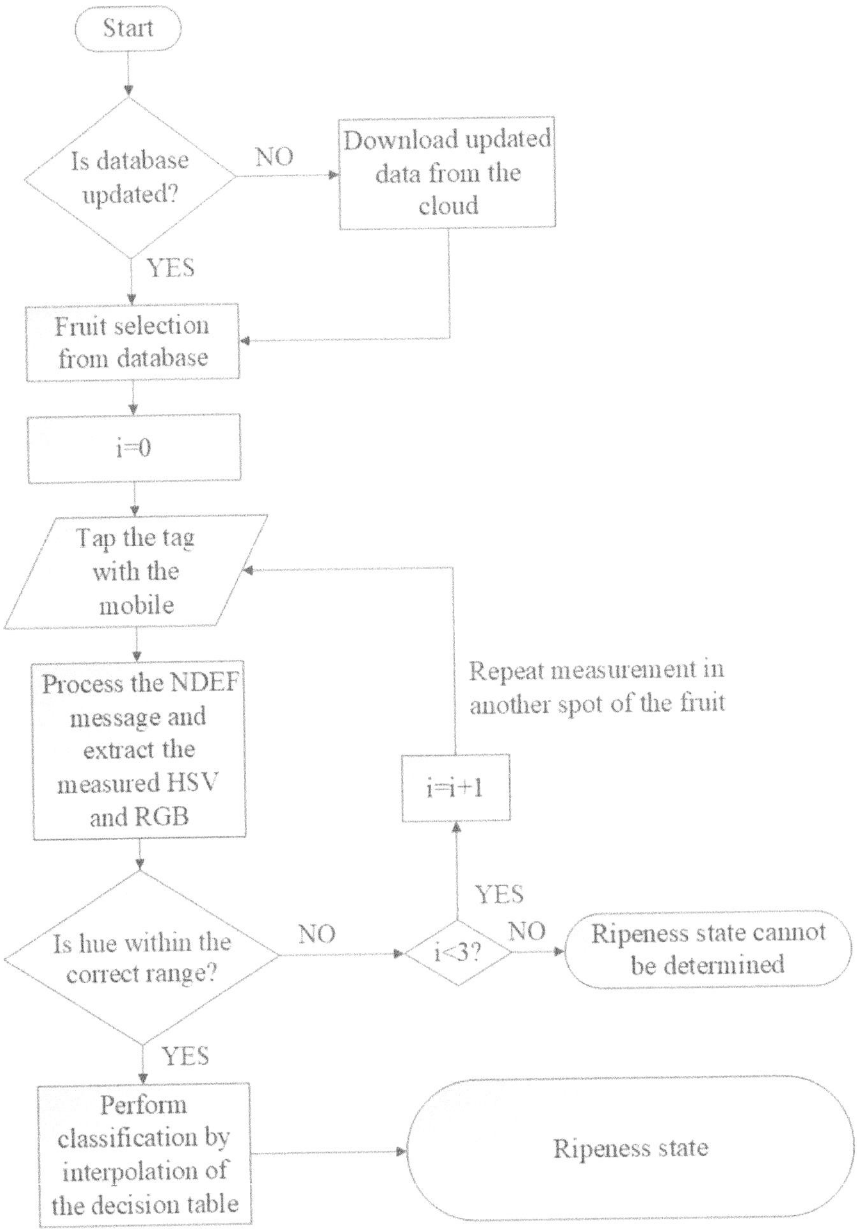

Figure 5.13. *Flowchart of the mobile application*

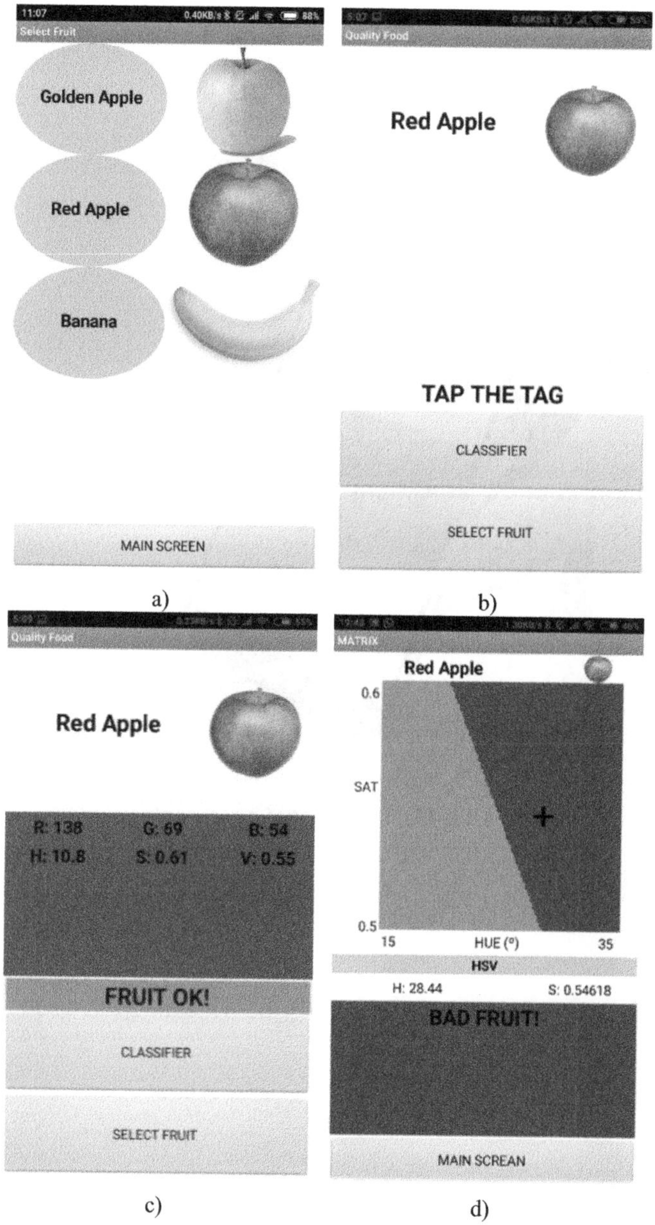

Figure 5.14. *Phone screen of the developed Android application. (a) Fruit selection; (b) screen indicating that the user can tap the tag; (c) representation of the detected color; and (d) decision boundaries of the training*

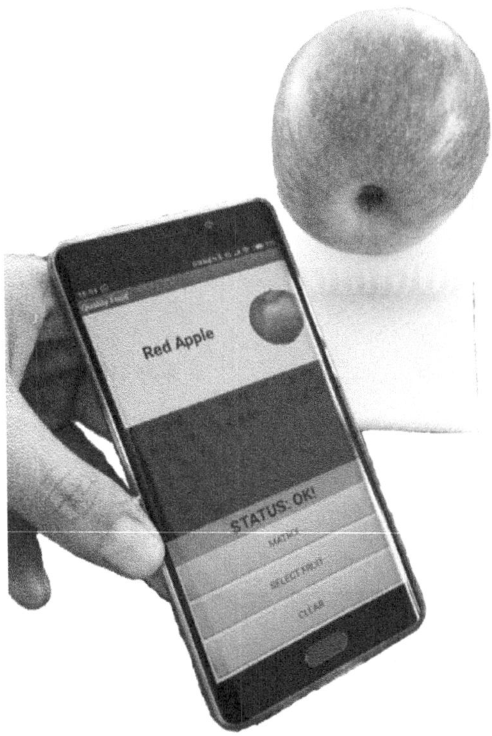

Figure 5.15. *Measurement of a red apple using the designed system and application*

List of Acronyms

AC	Alternating Current
ADC	Analog-to-Digital Converter
AIDC	Automatic Identification Data Capture
ASCII	American Standard Code for Information Interchange
ASIC	Application Specific Integrated Circuit
ASK	Amplitude Shift Keying
BLE	Bluetooth Low Energy
BPF	Band-Pass Filter
CDF	Cumulative Distribution Function
CMOS	Complementary Metal Oxide Semiconductor
CMT	Coupled-Mode Theory
CTI	Critical Temperature Indicator
DBS	Deep Brain Stimulation
DC	Direct Current
ECL	ElectroChemiLuminescence

EEPROM	Electrically Erasable Programmable Read-Only Memory
EH	Energy Harvesting
EMI	ElectroMagnetic Interference
FDR	Frequency-Domain Reflectometry
FR4	Fiberglass substrate
GPIO	General Purpose Input/Output
HSV	Hue, Saturation, Value
I^2C/I2C	Inter-Integrated Circuit
IC	Integrated Circuit
IDC	Inter-Digital Capacitor
IDE	Integrated Development Environment
IEC	International Electrotechnical Commission
IMD	Implantable Medical Device
IoT	Internet of Things
ISO	International Organization for Standardization
LDA	Linear Discrimination Analysis
MCU	MicroController Unit
MEMS	Micro-Electro-Mechanical Systems
MOS	Metal Oxide Semiconductor
MUT	Material Under Test
MWCNT	Multi-Walled Carbon NanoTubes
NDEF	NFC Data Exchange Format

NFC	Near-Field Communication
NMM	Neutron Moisture Meters
NMOS	N-type Metal Oxide Semiconductor
NN	Nearest Neighbor
P2P	Peer to Peer
PCA	Principal Component Analysis
PCB	Printed Circuit Board
PCE	Power Conversion Efficiency
PDL	Power Delivered to the Load
PLA	PolyLactic Acid
PTE	Power Transfer Efficiency
QDA	Quadratic Discriminant Analysis
QR	Quick Response
RF	Radio Frequency
RFID	Radio-Frequency IDentification
RGB	Red, Green, Blue
RX	Reception
SA	Spectrum Analyzer
SAP	SuperAbsorbent Polymer
SMD	Surface-Mount Device
TDR	Time-Domain Reflectometry
TVBN	Total Volatile Basic Nitrogen

TX	Transmission
UL	UltraLam
USB	Universal Serial Bus
VNA	Vector Network Analyzer
VWC	Volumetric Water Content
WISP	Wireless Identification and Sensing Platform
WLC	WireLess Charging
WPT	Wireless Power Transfer

References

Abad, E., Palacio, F., Nuin, M., de Zárate, A.G., Juarros, A., Gómez, J.M., Marco, S. (2009). RFID smart tag for traceability and cold chain monitoring of foods: Demonstration in an intercontinental fresh fish logistic chain. *Journal of Food Engineering*, 93(4), 394–399.

Adam, R. (2008). Skin care of the diaper area. *Pediatric Dermatology*, 25(4), 427–433.

Akin, T., Najafi, K., Bradley, R.M. (1998). A wireless implantable multichannel digital neural recording system for a micromachined sieve electrode. *IEEE Journal of Solid-State Circuits*, 33(1), 109–118.

Akkoç, F. and Şahin, Ö. (2016). Development and experimental comparison of low-cost, reliable capacitive touch sensing boards. *Sensors*, 16(11), 1786.

Alabdulhafith, M., Sampangi, R.V., Sampalli, S. (2013). NFC-enabled smartphone application for drug interaction and drug allergy detection. *Proceedings of the 5th International Workshop on Near Field Communication*, IEEE, 1–6, Zurich, Switzerland.

Alcalde, F., Alpago, O., Lipovetsky, J. (2014). CMOS design of an RFID interface compatible with ISO/IEC-14443 type a protocol. *Proceedings of the 2014 Argentine Conference on Micro-Nanoelectronics, Technology and Applications*, 41–47, Mendoza, Argentina.

Altawy, R. and Youssef, A.M. (2016). Security tradeoffs in cyber physical systems: A case study survey on implantable medical devices. *IEEE Access*, 4, 959–979.

AMS (2013). AS3953 – 14443 High Speed Passive Tag Interface. *ams AG* [Online]. Available at: https://ams.com/documents/20143/36005/AS3953_PB000100_3-00.pdf/07e0ef0a-407f-e77a-acc9-dc929e3a1353.

AMS (2014). SL13A – Smart Sensory Tag Chip For Unique Identification, Monitoring and Data Logging. *ams AG* [Online]. Available at: https://www.mouser.com/datasheet/2/588/AMS_SL13A_Datasheet_EN_v4-371531.pdf.

AMS (2017). AS3955 – NFC Forum Compliant Dynamic Tag. *ams AG* [Online]. Available at: https://ams.com/documents/20143/36005/AS3955_DS000400_6-00.pdf/4eed4b57-8384-9bbb-34d4-3e1125f7480f.

Anabtawi, N., Freeman, S., Ferzli, R. (2016). A fully implantable, NFC enabled, continuous interstitial glucose monitor. *Proceedings of the 2016 IEEE-EMBS International Conference on Biomedical and Health Informatics*, IEEE, 612–615, Nevada, USA.

Andrew, A.M. (2001). An introduction to support vector machines and other kernel-based learning methods. *Kybernetes*, 30(1), 103–115.

Ang, L.M., Ow, S.H., Seng, K.P., Tee, Z.H., Lee, B.W., Thong, M.K., Poi, P.J.H., Kunanayagam, S. (2008). Wireless intelligent incontinence management system using smart diapers. *Proceedings of the 5th International Conference on Electrical Engineering/Electronics, Computer, Telecommunications and Information Technology*, 69–72, Krabi, Thailand.

Atmel (2014). Atmel 8-bit AVR microcontroller with 2/4/8K bytes in-system programmable flash – ATtiny25/V, ATtiny45/V, ATtiny85/V [Online]. Available at: https://ww1.microchip.com/downloads/en/DeviceDoc/Atmel-2586-AVR-8-bit-Microcontroller-ATtiny25-ATtiny45-ATtiny85_Datasheet.pdf.

Azzarelli, J.M., Mirica, K.A., Ravnsbæk, J.B., Swager, T.M. (2014). Wireless gas detection with a smartphone via RF communication. *Proceedings of the National Academy of Sciences*, 111(51), 18162–18166.

Bazaka, K. and Jacob, M. (2012). Implantable devices: Issues and challenges. *Electronics*, 2(4), 1–34.

Berg, R.W., Milligan, M.C., Sarbaugh, F.C. (1994). Association of skin wetness and pH with diaper dermatitis. *Pediatric Dermatology*, 11(1), 18–20.

Blake, K. and Bible, S. (2005). Measuring small changes in capacitive sensors. Application Note AN1014, Microchip Technology Inc. [Online]. Available at: http://educypedia.karadimov.info/library/01014a.pdf.

Blöckner, M., Danti, S., Forrai, J., Broll, G., De Luca, A. (2009). Please touch the exhibits! Using NFC-based interaction for exploring a museum. *Proceedings of the 11th International Conference on Human–Computer Interaction with Mobile Devices and Services*, New York, USA.

Boada, M., Lazaro, A., Villarino, R., Girbau, D. (2018a). Battery-less soil moisture measurement system based on a NFC device with energy harvesting capability. *IEEE Sensors Journal*, 18(13), 5541–5549. doi: 10.1109/JSEN.2018.2837388.

Boada, M., Lazaro, A., Villarino, R., Gil, E., Girbau, D. (2018b). Near-field soil moisture sensor with energy harvesting capability. *2018 48th European Microwave Conference, EuMC 2018*, 235–238.

Boada, M., Lazaro, A., Villarino, R., Girbau, D (2019). Battery-less NFC sensor for pH monitoring. *IEEE Access*, 7, 33226–33239. doi: 10.1109/ACCESS.2019.2904109.

Boes, K., Borde, L., Egger, R. (2015). The acceptance of NFC smart posters in tourism. *Proceedings of the 2015 International Conference on Information and Communication Technologies in Tourism*, Lugano, Switzerland.

Breiman, L., Friedman, J.H., Olshen, R.A., Stone, C.J. (2017). *Classification and Regression Trees*. CRC Press, Boca Raton.

Bremner, J.M. and Mulvaney, C.S. (1982). Nitrogen – Total. In *Methods of Soil Analysis: Part 2, Chemical and Microbiological Properties*, Page, A.L. (ed.). American Society of Agronomy, Inc., Madison.

Brown, W.C. (1984). The history of power transmission by radio waves. *IEEE Transactions on Microwave Theory and Techniques*, 32(9), 1230–1242.

Brown, W.C. (1996). The history of wireless power transmission. *Solar Energy*, 56(1), 3–21.

Buchwald, J.Z. (1994). *The Creation of Scientific Effects: Heinrich Hertz and Electric Waves*. University of Chicago Press, Chicago.

Campbell Scientific (2017). TDR200 – Time-Domain Reflectometer [Online]. Available at: https://www.campbellsci.com/tdr200.

Cao, Z., Chen, P., Ma, Z., Li, S., Gao, X., Wu, R., Pan, L., Shi, Y. (2019). Near-field communication sensors. *Sensors*, 19(18), 3947.

Capel-Cuevas, S., López-Ruiz, N., Martinez-Olmos, A., Cuéllar, M.P., del Carmen Pegalajar, M., Palma, A.J., de Orbe-Payá, I., Capitán-Vallvey, L.F. (2012). A compact optical instrument with artificial neural network for pH determination. *Sensors*, 12(5), 6746–6763.

Cardullo, M.W. (1973). US3713148A – Transponder apparatus and system [Online]. Available at: https://patents.google.com/patent/US3713148A/en.

Cataldo, A., De Benedetto, E., Cannazza, G., Piuzzi, E., Pittella, E. (2018). TDR-based measurements of water content in construction materials for in-the-field use and calibration. *IEEE Transactions on Instrumentation and Measurement*, 67(5), 1230–1237.

Cerruela García, G., Luque Ruiz, I., Gómez-Nieto, M. (2016). State of the art, trends and future of bluetooth low energy, near field communication and visible light communication in the development of smart cities. *Sensors*, 16(11), 1968.

Chen, C. and Dong, T. (2015). Microfluidic paper-based analytical devices for colorimetric detection of urinary tract infection biomarkers on adult diapers. *Proceedings of the 37th Annual International Conference of the IEEE Engineering in Medicine and Biology Society*, 5892–5895, Milan, Italy.

Chen, H. and Zhao, A. (2016). NFC antenna for portable device with metal back cover. *Proceedings of the 2016 IEEE Antennas and Propagation Society International Symposium*, IEEE, 1471–1472, Farjado, Puerto Rico.

Chen, J.Z., Darhuber, A.A., Troian, S.M., Wagner, S. (2004). Capacitive sensing of droplets for microfluidic devices based on thermocapillary actuation. *Lab on a Chip*, 4(5), 473.

Chen, W., Shi, Y.Q., Xuan, G. (2007). Identifying computer graphics using HSV color model and statistical moments of characteristic functions. *Proceedings of the 2007 IEEE International Conference on Multimedia and Expo*, IEEE, 1123–1126, Beijing, China.

Chen, C., Wu, Y., Dong, T. (2014). Dipsticks integrated on smart diapers for colorimetric analysis of urinary tract infections in the field. *Proceedings of the 16th International Conference on Mechatronics*, 423–427, Brno, Czech Republic.

Chow, H.K.H., Choy, K.L., Lee, W.B., Lau, K.C. (2006). Design of a RFID case-based resource management system for warehouse operations. *Expert Systems with Applications*, 30(4), 561–576.

Chumkamon, S., Tuvaphanthaphiphat, P., Keeratiwintakorn, P. (2008). A blind navigation system using RFID for indoor environments. *Proceedings of the 5th International Conference on Electrical Engineering/Electronics, Computer, Telecommunications and Information Technology*, IEEE, 2, 765–768, Krabi, Thailand.

Chung, M.A. and Yang, C.F. (2016). Miniaturized NFC antenna design for a tablet PC with a narrow border and metal back-cover. *IEEE Antennas and Wireless Propagation Letters*, 15, 1470–1474.

Cohn, S.B. (1958). Parallel-coupled transmission-line-resonator filters. *IEEE Transactions on Microwave Theory and Techniques*, 6(2), 223–231.

Coskun, V., Ozdenizci, B., Ok, K. (2015). The survey on near field communication. *Sensors*, 15(6), 13348–13405.

Cui, L., Zhang, Z., Gao, N., Meng, Z., Li, Z. (2019). Radio frequency identification and sensing techniques and their applications – A review of the state-of-the-art. *Sensors*, 19(18), 4012.

De Maeyer, J., Devos, H., Meeus, W., Verplaetse, P., Stroobandt, D. (2003). Hardware implementation of an EAN-13 bar code decoder. *Proceedings of the 2003 Asia and South Pacific Design Automation Conference*, IEEE, 583–584, Kitakyushu, Japan.

Dean, T.J., Bell, J.P., Baty, A.J.B. (1987). Soil moisture measurement by an improved capacitance technique, Part I: Sensor design and performance. *Journal of Hydrology*, 93(1–2), 67–78.

DeHennis, A., Getzlaff, S., Grice, D., Mailand, M. (2016). An NFC-enabled CMOS IC for a wireless fully implantable glucose sensor. *IEEE Journal of Biomedical and Health Informatics*, 20(1), 18–28.

Dominguez, R.B., Orozco, M.A., Chávez, G., Márquez-Lucero, A. (2017). The evaluation of a low-cost colorimeter for glucose detection in salivary samples. *Sensors*, 17(11), 19–21.

Douthwaite, M. and Georgiou, P. (2017). Live demonstration: An NFC based batteryless CMOS ISFET array for real-time pH measurements of bio-fluids. *Proceedings of IEEE Sensors 2017*, Glasgow, Scotland.

Drach, G.W., Layton, T.N., Binard, W.J. (1979). Male peak urinary flow rate: Relationships to volume voided and age. *The Journal of Urology*, 122(2), 210–214.

EM Microelectronics (2014). NF4 – NFC forum type 4 dual interface [Online]. Available at: http://www.advanide.com/wp-content/uploads/products/rfid/nf4-fs.pdf.

Ervasti, M., Isomursu, M., Kinnula, M. (2009). Experiences from NFC supported school attendance supervision for children. *Proceedings of the 3rd International Conference on Mobile Ubiquitous Computing, Systems, Services, and Technologies*, IEEE, 22–30, Silema, Malta.

Escobedo, P., Erenas, M.M., López-Ruiz, N., Carvajal, M.A., Gonzalez-Chocano, S., de Orbe-Payá, I., Capitán-Valley, L.F., Palma, A.J., Martínez-Olmos, A. (2017). Flexible passive near field communication tag for multigas sensing. *Analytical Chemistry*, 89(3), 1697–1703.

Escobedo, P., Erenas, M.M., Martínez-Olmos, A., Carvajal, M.A., Gonzalez-Chocano, S., Capitán-Vallvey, L.F., Palma, A.J. (2019). General-purpose passive wireless point-of-care platform based on smartphone. *Biosensors and Bioelectronics*, 141, 111360.

Evett, S.R., Tolk, J.A., Howell, T.A. (2003). A depth control stand for improved accuracy with the neutron probe. *Vadose Zone Journal*, 2(4), 642–649.

Falas, T. and Kashani, H. (2007). Two-dimensional bar-code decoding with camera-equipped mobile phones. *Proceedings of the Fifth Annual IEEE International Conference on Pervasive Computing and Communications Workshops*, IEEE, 597–600, White Plains, USA.

Finkenzeller, K. (2010). *RFID Handbook: Fundamentals and Applications in Contactless Smart Cards, Radio Frequency Identification and Near-Field Communication*. John Wiley & Sons, Ltd, Chichester.

Fischbach, F.T. and Dunning, M.B. (2009). *A Manual of Laboratory and Diagnostic Tests*. Wolters Kluwer Health/Lippincott Williams & Wilkins, Philadelphia, USA.

Flack, F.C., James, E.D., Schlapp, D.M. (1971) Mutual inductance of air-cored coils: Effect on design of radio-frequency coupled implants. *Medical & Biological Engineering*, 9(2), 79–85.

Fontecha, J., Hervás, R., Bravo, J., Villarreal, V. (2011). An NFC approach for nursing care training. *Proceedings of the 3rd International Workshop on Near Field Communication,* 38–43, Hagenberg, Austria.

Fressancourt, A., Hérault, C., Ptak, E. (2009). NFCSocial: Social networking in mobility through IMS and NFC. *Proceedings of the 1st International Workshop on Near Field Communication*, IEEE, 24–29, Hagenberg, Austria.

Gabriel, S., Lau, R.W., Gabriel, C. (1996a). The dielectric properties of biological tissues II: Measurements in the frequency range 10 Hz to 20 GHz. *Physics in Medicine and Biology*, 41(11), 2251–2269.

Gabriel, S., Lau, R.W., Gabriel, C. (1996b). The dielectric properties of biological tissues III: Parametric models for the dielectric spectrum of tissues. *Physics in Medicine and Biology*, 41(41), 2251–2269.

Gaitán-Pitre, J.E., Gasulla, M., Pallàs-Areny, R. (2009). Analysis of a direct interface circuit for capacitive sensors. *IEEE Transactions on Instrumentation and Measurement*, 58(9), 2931–2937.

García-García, J., Martín, F., Falcone, F., Bonache, J., Gil, I., Lopetegi, T., Laso, M.A.G., Sorolla, M., Marqués, R. (2004). Spurious passband suppression in microstrip coupled line band pass filters by means of split ring resonators. *IEEE Microwave and Wireless Components Letters*, 14(9), 416–418.

Gebhart, M. (2011). Analytical considerations for an ISO/IEC14443 compliant SmartCard transponder. *Proceedings of the 11th International Conference on Telecommunications*, IEEE, 9–16, Graz, Austria.

Gebhart, M., Bruckbauer, J., Gossar, M. (2010). Chip impedance characterization for contactless proximity personal cards. *Proceedings of the 7th International Symposium on Communication Systems, Networks and Digital Signal Processing*, IEEE, 826–830, Newcastle upon Tyne, UK.

Gevorgian, S. and Berg, H. (2001). Line capacitance and impedance of coplanar-strip waveguides on substrates with multiple dielectric layers. *Proceedings of the 31st European Microwave Conference*, IEEE, 1–4, London UK.

Gevorgian, S.S., Martinsson, T., Linnér, P.L.J., Kollberg, E.L. (1996). CAD models for multilayered substrate interdigital capacitors. *IEEE Transactions on Microwave Theory and Techniques*, 44(6), 896–904.

Ghìron, S.L., Sposato, S., Medaglia, C.M., Moroni, A. (2009). NFC ticketing: A prototype and usability test of an NFC-based virtual ticketing application. *Proceedings of the 1st International Workshop on Near Field Communication*, 45–50, Hagenberg, Austria.

Giantec Semiconductor (2011). GT23SC6699-1/2 NFC Type 2 [Online]. Available at: http://www.giantec-semi.com/Upload/datasheet/GT23SC6699_DS_ADV.pdf.

Gutierrez, J., Villa-Medina, J.F., Nieto-Garibay, A., Porta-Gandara, M.A. (2014). Automated irrigation system using a wireless sensor network and GPRS module. *IEEE Transactions on Instrumentation and Measurement*, 63(1), 166–176.

Gvozdenovic, N., Mayer, L.W., Mecklenbrauker, C.F. (2014). Measurement of harmonic distortions and impedance of HF RFID chips. *Proceedings of the 8th European Conference on Antennas and Propagation*, 2940–2944, The Hague, The Netherlands.

Hahnel, D., Burgard, W., Fox, D., Fishkin, K., Philipose, M. (2004). Mapping and localization with RFID technology. *Proceedings of the IEEE International Conference on Robotics and Automation*, IEEE, 1015–1020, New Orleans, USA.

Hamam, R.E., Karalis, A., Joannopoulos, J.D., Soljačić, M. (2007). Coupled-mode theory for general free-space resonant scattering of waves. *Physical Review A – Atomic, Molecular, and Optical Physics*, 75(5), 1–5.

Han, S., Kim, J., Won, S.M., Ma, Y., Kang, D., Xie, Z., Lee, K.-T., Chung, H.U., Banks, A., Min, S., Heo, S.Y., Davies, C.R., Lee, J.W., Lee, C.-H., Kim, B.H., Li, K., Zhou, Y., Wei, C., Feng, X., Huang, Y., Rogers, J.A. (2018). Battery-free, wireless sensors for full-body pressure and temperature mapping. *Science Translational Medicine*, 10(435), eaan4950.

Hannan, M.A., Mutashar, S., Samad, S.A., Hussain, A. (2014). Energy harvesting for the implantable biomedical devices: Issues and challenges. *BioMedical Engineering Online*, 13(1), 1–23.

Hashimoto, Y., Hamaguchi, M., Nakanishi, N., Ohbora, A., Kojima, T., Fukui, M. (2017). Urinary pH is a predictor of diabetes in men: A population-based large scale cohort study. *Diabetes Research and Clinical Practice*, 130, 9–14.

Honeywell (2001). HIH-5030/5031 series – Low voltage humidity sensors [Online]. Available at: https://datasheet.octopart.com/HIH-5030-001-Honeywell-datasheet-11765261.pdf.

Hu, J., Wu, S., Lin, K., Tan, H. (2009). Low-power energy supply circuit for passive RFID transponder. *Proceedings of the 2009 IEEE International Conference on RFID*, 1–6, Orlando, USA.

Huang, S.Y. (2010). The effects of ground planes in modern electromagnetic applications. PhD thesis, Nanyang Technological University, Singapore.

Huang, C.H. and Chang, S.L. (2013). Study on the feasibility of NFC P2P communication for nursing care daily work. *Journal of Computers (Taiwan)*, 24(2), 33–45.

HunterLab (2008). Colorimeters versus spectrophotometers [Online]. Available at: https://pdf4pro.com/view/colorimeters-versus-spectrophotometers-1b2a8a.html.

Ishii, M. and Komiyama, K. (2004). A measurement method for the antenna factor of small loop antenna by measuring the input impedance. *Proceedings of the 2004 Conference on Precision Electromagnetic Measurements*, IEEE, 80–81, London, UK.

ISO/IEC (2010). ISO/IEC 15693: Identification cards – Contactless integrated circuit cards – Vicinity cards. International Organization for Standardization. Geneva, Switzerland.

ISO/IEC (2013). ISO/IEC 18092: Information technology – Telecommunications and information exchange between systems – Near field communication – Interface and Protocol (NFCIP-1). International Organization for Standardization. Geneva, Switzerland.

ISO/IEC (2018). ISO/IEC 14443: Identification cards – Contactless integrated circuit cards – Proximity cards. International Organization for Standardization. Geneva, Switzerland.

ISO/IEC (2000a). ISO/IEC 16022: Information technology – International symbology specification – Data Matrix. International Organization for Standardization. Geneva, Switzerland.

ISO/IEC (2000b). ISO/IEC 18004: Information technology – Automatic identification and data capture techniques – Bar code symbology – QR Code. International Organization for Standardization. Geneva, Switzerland.

Jacobi, R. and La Cost, E. (2017). Antenna design guide for the TRF79xxA application report [Online]. Available at: www.ti.com.

Jamali, B., Ranasinghe, D.C., Cole, P.H. (2005). Analysis of UHF RFID CMOS rectifier structures and input impedance characteristics. *Microelectronics: Design, Technology, and Packaging II*, 6035(January), 60350D.

Jeong, H., Ha, T., Kuang, I., Shen, L., Dai, Z., Sun, N., Lu, N. (2017). NFC-enabled, tattoo-like stretchable biosensor manufactured by "cut-and-paste" method. *Proceedings of the 39th Annual International Conference of the IEEE Engineering in Medicine and Biology Society*, IEEE, 4094–4097, Jeju Island, South Korea.

Jia, X., Feng, Q., Fan, T., Lei, Q. (2012). RFID technology and its applications in Internet of Things (IoT). *Proceedings of the 2nd International Conference on Consumer Electronics, Communications and Networks*, IEEE, 1282–1285, Three Gorges, China.

Jiang, B., Smith, J.R., Philipose, M., Roy, S., Sundara-Rajan, K., Mamishev, A.V. (2005). Energy scavenging for inductively coupled passive RFID systems. *Proceedings of the 2005 Instrumentationand Measurement Technology Conference*, IEEE, 984–989, Ottawa, Canada.

Jianguo, H., Hongzhou, T., Yongjian, X. (2007). A novel design of low power RFID tag interface circuit. *IET Conference Publications*, (533 CP), 478–481.

Jones, S.B., Wraith, J.M., Or, D. (2002). Time domain reflectometry measurement principles and applications. *Hydrological Processes*, 16(1), 141–153.

Juels, A. (2006). RFID security and privacy: A research survey. *IEEE Journal on Selected Areas in Communications*, 24(2), 381–394.

Karlsen, H. and Dong, T. (2015). A smart phone-based robust correction algorithm for the colorimetric detection of Urinary Tract Infection. *Proceedings of the Annual International Conference of the IEEE Engineering in Medicine and Biology Society*, IEEE, 1251–1254, New Orleans, USA.

Karpischek, S., Michahelles, F., Resatsch, F., Fleisch, E. (2009). Mobile sales assistant: An NFC-based product information system for retailers. *Proceedings of the 1st International Workshop on Near Field Communication*, IEEE, 20–23, Hagenberg, Austria.

Kassal, P., Steinberg, I.M., Steinberg, M.D. (2013). Wireless smart tag with potentiometric input for ultra low-power chemical sensing. *Sensors and Actuators B: Chemical*, 184, 254–259.

Ke, X., Guandong, G., Jian, L. (2013). An improved method of detecting pork freshness based on CRR features. *Proceedings of the 9th International Conference on Intelligent Information Hiding and Multimedia Signal Processing*, IEEE, 194–197, Beijing, China.

Kelly, E.P. and Erickson, G.S. (2005). RFID tags: Commercial applications v. privacy rights. *Industrial Management and Data Systems*, 105(6), 703–713.

Khan, T. (2018). A smart wearable gadget for noninvasive detection and notification of diaper moister. *Proceedings of the 2018 IEEE International Conference on Electro Information Technology*, IEEE, 240–244, Lincoln, USA.

Kieseberg, P., Leithner, M., Mulazzani, M., Munroe, L., Schrittwieser, S., Sinha, M., Weippl, E. (2010). QR code security. *Proceedings of the 8th International Conference on Advances in Mobile Computing and Multimedia*, 430, Paris, France.

Kifle, Y., Wikner, J., Zotterman, J., Ryden, L., Farnebo, S. (2019). NFC powered implantable temperature sensor. *Proceedings of the 41st Annual International Conference of the IEEE Engineering in Medicine and Biology Society*, IEEE, 4359–4362, Berlin, Germany.

Kim, J.W. (2008). *Development of Interdigitated Capacitor Sensors for Direct and Wireless Measurements of the Dielectric Properties of Liquids*. University of Texas Libraries, Austin.

Kim, K.Y. (ed.) (2012). *Wireless Power Transfer – Principles and Engineering Explorations*. IntechOpen, London.

Kim, J., Banks, A., Xie, Z., Heo, S.Y., Gutruf, P., Lee, J.W., Xu, S., Jang, K.I., Liu, F., Brown, G., Choi, J., Kim, J.H., Feng, X., Huang, Y., Paik, U., Rogers, J.A. (2015). Miniaturized flexible electronic systems with wireless power and near-field communication capabilities. *Advanced Functional Materials*, 25(30), 4761–4767.

Kim, S.D., Koo, Y., Yun, Y. (2017). A smartphone-based automatic measurement method for colorimetric pH detection using a color adaptation algorithm. *Sensors*, 17(7), 1604.

Klute, A. (1986). *Methods of Soil Analysis: Part 1, Physical and Mineralogical Methods*. American Society of Agronomy, Madison, USA.

Koh, A., Kang, D., Xue, Y., Lee, S., Pielak, R.M., Kim, J., Hwang, T., Min, S., Banks, A., Bastien, P., Manco, M.C., Wang, L., Ammann, K.R., Jang, K.I., Won, P., Han, S., Ghaffari, R., Paik, U., Slepian, M.J., Balooch, G., Huang, Y., Rogers, J.A. (2016). A soft, wearable microfluidic device for the capture, storage, and colorimetric sensing of sweat. *Science Translational Medicine*, 8(366), 1–14.

Kollegger, C., Greiner, P., Steffan, C., Wiessflecker, M., Froehlich, H., Kautzsch, T., Holweg, G., Deutschmann, B. (2017). A system-on-chip NFC bicycle tire pressure measurement system. *Proceedings of the 60th International Midwest Symposium on Circuits and Systems*, IEEE, 60–63, Boston, USA.

Kosemund, K., Schlatter, H., Ochsenhirt, J.L., Krause, E.L., Marsman, D.S., Erasala, G.N. (2009). Safety evaluation of superabsorbent baby diapers. *Regulatory Toxicology and Pharmacology*, 53(2), 81–89.

Köstinger, H., Gobber, M., Grechenig, T., Tappeiner, B., Schramm, W. (2013). Developing a NFC based patient identification and ward round system for mobile devices using the android platform. *Proceedings of the 2013 Special Topic Conference on Point-of-Care (POC) Healthcare Technologies: Synergy Towards Better Global Healthcare*, IEEE, 176–179, Bangalore, India.

Kurs, A., Karalis, A., Moffatt, R., Joannopoulos, J.D., Fisher, P., Soljacic, M. (2007). Wireless power transfer via strongly coupled magnetic resonances. *Science*, 317(5834), 83–86.

Lahtela, A., Hassinen, M., Jylhä, V. (2008). RFID and NFC in healthcare: Safety of hospitals medication care. *Proceedings of the 2nd International Conference on Pervasive Computing Technologies for Healthcare*, 241–244, Tampere, Finland.

Lazaro, A., Villarino, R., Girbau, D. (2018). A survey of NFC sensors based on energy harvesting for IoT applications. *Sensors*, 18(11), 3746.

Lazaro, A., Boada, M., Villarino, R., Girbau, D. (2019). Battery-less smart diaper based on NFC technology. *IEEE Sensors Journal*, 19(22), 10848–10858. doi: 10.1109/JSEN.2019.2933289.

Lee, D.J., Archibald, J.K., Chang, Y.C., Greco, C.R. (2008). Robust color space conversion and color distribution analysis techniques for date maturity evaluation. *Journal of Food Engineering*, 88(3), 364–372.

Lee, D.J., Archibald, J.K., Xiong, G. (2011). Rapid color grading for fruit quality evaluation using direct color mapping. *IEEE Transactions on Automation Science and Engineering*, 8(2), 292–302.

Lee, B., Kim, B., Yang, S. (2014). Enhanced loop structure of NFC antenna for mobile handset applications. *International Journal of Antennas and Propagation*, 2014, 1–6.

Li, G.B., Ke, J.M., Zhang, H.J. (2015). The influence on the absorption properties of paper diapers by the SAP mass ratio and fluff pulp. *Advanced Materials Research*, 1089, 33–36.

Liao, K.-C. and Lee, W.-H. (2010). A novel user authentication scheme based on QR-Code. *Journal of Networks*, 5(8), 937–941.

Lin, Y.P., Yeh, C.Y., Huang, P.Y., Wang, Z.Y., Cheng, H.H., Li, Y.T., Chuang, C.F., Huang, P.C., Tang, K.T., Ma, H.P., Chang, Y.C., Yeh, S.R., Chen, H. (2016). A battery-less, implantable neuro-electronic interface for studying the mechanisms of deep brain stimulation in rat models. *IEEE Transactions on Biomedical Circuits and Systems*, 10(1), 98–112.

Liu, Y., Yang, J., Liu, M. (2008). Recognition of QR code with mobile phones. *Proceedings of the 2008 Chinese Control and Decision Conference*, IEEE, 203–206, Yantai, China.

Loh, B.Y., Vuong, N.K., Chan, S., Lau The, C.T. (2011). Robust classification of pH levels on a camera phone. *Proceedings of the 2011 International Multiconference of Engineers and Computer Scientists 2011*, 1, 600–604, Hong Kong.

Lopez-Ruiz, N., Curto, V.F., Erenas, M.M., Benito-Lopez, F., Diamond, D., Palma, A.J., Capitan-Vallvey, L.F. (2014). Smartphone-based simultaneous pH and nitrite colorimetric determination for paper microfluidic devices. *Analytical Chemistry*, 86(19), 9554–9562.

Lorite, G.S., Selkälä, T., Sipola, T., Palenzuela, J., Jubete, E., Viñuales, A., Cabañero, G., Grande, H.J., Tuominen, J., Uusitalo, S., Hakalahti, L., Kordas, K., Toth, G. (2017). Novel, smart and RFID assisted critical temperature indicator for supply chain monitoring. *Journal of Food Engineering*, 193, 20–28.

Lu, C.H., Li, J.A., Lin, T.H. (2016). A 13.56-MHz passive NFC tag IC in 0.18-μm CMOS process for biomedical applications. *Proceedings of the 2016 International Symposium on VLSI Design, Automation and Test*, Hsinchu, Taiwan.

Ma, Z., Chen, P., Cheng, W., Yan, K., Pan, L., Shi, Y., Yu, G. (2018). Highly sensitive, printable nanostructured conductive polymer wireless sensor for food spoilage detection. *Nano Letters*, 18(7), 4570–4575.

Macias, M., Bernabeu-Andreu, F.A., Arribas, I., Navarro, F., Baldominos, G. (2018). Impact of a barcode medication administration system on patient safety. *Oncology Nursing Forum*, 45(1), E1–E13.

Mainetti, L., Patrono, L., Vergallo, R. (2012). IDA-Pay: A secure and efficient micro-payment system based on Peer-to-Peer NFC technology for Android mobile devices. *Journal of Communications Software and Systems*, 8(4), 117–125.

Marechal, C. and Paret, D. (2008). The loading effect of proximity contactless smart card. Incidences of impedance of the shunt regulator. *Proceedings of the 2008 International Conference on Wireless Communications, Networking and Mobile Computing*, 8–11, Dalian, China.

Martinez, A.W., Phillips, S.T., Whitesides, G.M., Carrilho, E. (2010). Diagnostics for the developing world: Microfluidic paper-based analytical devices. *Analytical Chemistry*, 82(1), 3–10.

McCarty, C.G. and Vitz, E. (2006). pH paradoxes: Demonstrating that it is not true that pH = -log[H+]. *Journal of Chemical Education*, 83(5), 752–757.

McKnight, M., Lin, F., Kausche, H., Ghosh, T., Bozkurt, A. (2015). Towards paper based diaper sensors. *Proceedings of the 2015 Biomedical Circuits and Systems Conference: Engineering for Healthy Minds and Able Bodies*, IEEE, 1–4, Atlanta, USA.

Medlicott, A., Semple, A., Thompson, A., Blackbourne, H. (1992). Measurement of colour changes in ripening bananas and mangoes by instrumental, chemical and visual assessments. *Tropical Agriculture*, 69(2), 161–166.

MELEXIS (2012). MLX90129 – 13.56MHz sensor tag/datalogger [Online]. Available at: https://www.melexis.com/-/media/files/documents/datasheets/mlx90129-datasheet-melexis.pdf.

Mendoza, F., Dejmek, P., Aguilera, J.M. (2006). Calibrated color measurements of agricultural foods using image analysis. *Postharvest Biology and Technology*, 41(3), 285–295.

Microchip Technology (2015). PIC16(L)F1703/7 – 14/20-Pin 8-Bit Advanced Analog Flash Microcontroller [Online]. Available at: http://ww1.microchip.com/downloads/en/DeviceDoc/40001722C.pdf.

Mika, S., Ratsch, G., Weston, J., Scholkopf, B., Muller, K.R. (1999). Fisher discriminant analysis with kernels. *Proceedings of the 1996 IEEE Signal Processing Society Workshop*, 41–48, Kyoto, Japan.

Morak, J., Schwarz, M., Hayn, D., Schreier, G. (2012). Feasibility of mHealth and Near Field Communication technology based medication adherence monitoring. *Proceedings of the 34th Annual International Conference of the IEEE Engineering in Medicine and Biology Society*, IEEE, 272–275, San Diego, USA.

Muja, M. and Lowe, D.G. (2009). Fast approximate nearest neighbors with automatic algorithm configuration. *Proceedings of the 4th International Conference on Computer Vision Theory and Applications*, 331–340, Lisbon, Portugal.

Müller, J., Borovskiy, V., Panchenko, O., Bog, A., Zeier, A. (2009). NFC at the workplace – Simplify enterprise work flows with NFC. *Proceedings of the 2009 IEEE International Conference on Industrial Engineering and Engineering Management*, IEEE, 1539–1542, Beijing, China.

Muniz, R., Junco, L., Otero, A. (1999). A robust software barcode reader using the Hough transform. *Proceedings of the 1999 International Conference on Information Intelligence and Systems*, IEEE, 313–319, Bethesda, USA.

Murphy, M.F. and Kay, J.D. (2004). Barcode identification for transfusion safety. *Current Opinion in Hematology*, 11(5), 334–338.

Mustafa, N.B.A., Fuad, N.A., Ahmed, S.K., Abidin, A.A.Z., Ali, Z., Wong, B.Y., Sharrif, Z.A.M. (2008). Image processing of an agriculture produce: Determination of size and ripeness of a banana. *Proceedings of the International Symposium on Information Technology 2008, Kuala Lumpur*, Malaysia.

Ni, L.M-S., Liu, Y., Lau, Y.C., Patil, A.P. (2003). LANDMARC: Indoor location sensing using active RFID. *Proceedings of the 1st International Conference on Pervasive Computing and Communications*, IEEE, 407–415, Fort Worth, USA.

Nilsson, H.-E., Sidén, J., Gulliksson, M., Siden, J., Gulliksson, M. (2011). An incontinence alarm solution utilizing RFID based sensor technology. *Proceedings of the 2011 International Conference on RFID-Technologies and Applications*, IEEE, 359–363, Sitges, Spain.

NXP Semiconductors (2015a). NT3H1101/NT3H1201 – NTAG I2C [Online]. Available at: https://www.mouser.com/datasheet/2/302/NT3H1101_1201-1127167.pdf.

NXP Semiconductors (2015b). NTAG213/215/216 – NFC Forum Type 2 Tag compliant IC with 144/504/888 bytes user memory [Online]. Available at: https://www.nxp.com/docs/en/data-sheet/NTAG213_215_216.pdf.

NXP Semiconductors (2018a). AN12549 – PN5180 Antenna design guide [Online]. Available at: https://www.nxp.com/docs/en/application-note/AN12549.pdf.

NXP Semiconductors (2018b). PN7120 – NFC controller with integrated firmware, supporting all NFC Forum modes [Online]. Available at: https://www.nxp.com/docs/en/data-sheet/PN7120.pdf.

NXP Semiconductors (2018c). NHS3100 – Temperature logger IC [Online]. Available at: https://www.digchip.com/datasheets/parts/datasheet/1019/NHS3100_A1_515-pdf.php.

NXP Semiconductors (2019). AN11755 – PN7150 Antenna design and matching guide [Online]. Available at: https://www.nxp.com/docs/en/application-note/AN11755.pdf.

Ohbuchi, E., Hanaizumi, H., Ah Hock, L. (2004). Barcode readers using the camera device in mobile phones. *Proceedings of the 2004 International Conference on Cyberworlds*, IEEE, 260–265, Tokyo, Japan.

Ok, K., Aydin, M., Coskun, V., Ozdenizci, B. (2010). Exploring underlying values of NFC applications. *Proceedings of the International Conference on Management Technology and Applications*, 12, 290–294, Libertad City, Ecuador.

de Oliveira Filho, J.I. and Do Prado Villarroel Zurita, M.E. (2017). Development of NFC TAG for temperature sensing of premature newborns in neonatal incubators. *Proceedings of the 2nd International Symposium on Instrumentation Systems, Circuits and Transducers: Chip on the Sands, Proceedings*, IEEE, 1–4.

Oncescu, V., O'Dell, D., Erickson, D. (2013). Smartphone based health accessory for colorimetric detection of biomarkers in sweat and saliva. *Lab on a Chip*, 13(16), 3232–3238.

Ong, J.B., You, Z., Mills-Beale, J., Tan, E.L., Pereles, B.D., Ong, K.G. (2008). A wireless, passive embedded sensor for real-time monitoring of water content in civil engineering materials. *IEEE Sensors Journal*, 8(12), 2053–2058.

Pal, D., Vanijja, V., Papasratorn, B. (2015). An empirical analysis towards the adoption of NFC mobile payment system by the end user. *Procedia Computer Science*, 69, 13–25.

Pala, Z. and Inanç, N. (2007). Smart parking applications using RFID technology. *Proceedings of the 1st Annual RFID Eurasia*. Istanbul, Turkey.

Palmer, R.C. (2007). *The Bar Code Book: A Comprehensive Guide to Reading, Printing, Specifying, and Applying Bar Code and Other Machine-Readable Symbols*. Trafford Publishing, Victoria, Canada.

Paret, D. (2016). *Antenna Designs for NFC Devices*. ISTE Ltd, London and John Wiley & Sons, New York, USA.

Pasquet, M., Reynaud, J., Rosenberger, C. (2008). Secure payment with NFC mobile phone in the SmartTouch project. *Proceedings of the 2008 International Symposium on Collaborative Technologies and Systems*, IEEE, 121–126, Irvine, USA.

Pathare, P.B., Opara, U.L., Al-Said, F.A.J. (2013). Colour measurement and analysis in fresh and processed foods: A review. *Food and Bioprocess Technology*, 6(1), 36–60.

Peng, L., Breinbjerg, O., Mortensen, N.A. (2010). Wireless energy transfer through non-resonant magnetic coupling. *Journal of Electromagnetic Waves and Applications*, 24(11–12), 1587–1598.

Pierre, F., Aujol, J.F., Bugeau, A., Ta, V.T. (2015). Luminance-hue specification in the RGB space. *Proceedings of the 5th International Conference on Scale Space and Variational Methods in Computer Vision*, 413–424, Lège-Cap Ferret, France.

Pourghomi, P., Qasim, M., Ghinea, G. (2013). A proposed NFC payment application. *International Journal of Advanced Computer Science and Applications*, 4(8), 173–181.

Prowle, J.R., Liu, Y.-L., Licari, E., Bagshaw, S.M., Egi, M., Haase, M., Haase-Fielitz, A., Kellum, J.A., Cruz, D., Ronco, C., Tsutsui, K., Uchino, S., Bellomo, R. (2011). Oliguria as predictive biomarker of acute kidney injury in critically ill patients. *Critical Care*, 15(4), R172.

Puma, J.P., Huerta, M., Alvizu, R., Clotet, R. (2012). Mobile identification: NFC in the healthcare sector. *Proceedings of the 6th Andean Region International Conference, Andescon*, 39–42, Cuenca, Ecuador.

Rahimi, R., Brener, U., Ochoa, M., Ziaie, B. (2017). Flexible and transparent pH monitoring system with NFC communication for wound monitoring applications. *Proceedings of the 30th International Conference on Micro Electro Mechanical Systems*, Las Vegas, USA.

Rahman, M.S., Choi, C., Kim, Y.P., Kim, S., Choi, J. (2017). A low-cost wet diaper detector based on smart phone and BLE sensor. *International Journal of Applied Engineering Research*, 12(19), 9074–9077.

Ramos de Luna, I., Montoro-Ríos, F., Liébana-Cabanillas, F. (2015). Determinants of the intention to use NFC technology as a payment system: An acceptance model approach. *Information Systems and e-Business Management*, 14(2), 293–314.

Recce, M., Taylor, J., Plebe, A., Tropiano, G. (1996). High speed vision-based quality grading of oranges. *Proceedings of the International Workshop on Neural Networks for Identification, Control, Robotics, and Signal/Image Processing*, IEEE, 136–144, Venice, Italy.

Rennick, G., Attikiouzel, Y., Zaknich, A. (1999). Machine grading and blemish detection in apples. *Proceedings of the 5th International Symposium on Signal Processing and Its Applications*, IEEE, 567–570 Brisbane, Australia.

Reynolds, S.G. (1970). The gravimetric method of soil moisture determination. Part I: A study of equipment, and methodological problems. *Journal of Hydrology*, 11(3), 258–273.

Rish, I. (2001). IBM research report – An empirical study of the naive Bayes classifier [Online]. Available at: https://dibernardo.tigem.it/files/papers/2001/rc22230.pdf/.

Robinson, D.A., Jones, S.B., Wraith, J.M., Or, D., Friedman, S.P. (2002). A review of advances in dielectric and electrical conductivity measurement in soils using time domain reflectometry. *Vadose Zone Journal*, 2(4), 444–475.

Rogers Corporation (2015). ULTRALAM® 3000 – Liquid crystalline polymer circuit material: Double-clad laminates [Online]. Available at: https://www.midwestpcb.com/data_sheets/RogersULTRALAM.pdf.

Rose, D.P., Ratterman, M., Griffin, D.K., Hou, L., Kelley-Loughnane, N., Naik, R.K., Hagen, J.A., Papautsky, I., Heikenfeld, J. (2014). System-level design of an RFID sweat electrolyte sensor patch. *Proceedings of the 36th Annual International Conference of the IEEE Engineering in Medicine and Biology Society*, 4038–4041, Chicago, USA.

Ruiz, I.L. and Gómez-Nieto, M.Á. (2009). University smart poster: Study of NFC technology applications for university ambient. *Advances in Soft Computing*, 51, 112–116.

Sajal, S., Atanasov, Y., Braaten, B.D., Marinov, V., Swenson, O. (2014). A low cost flexible passive UHF RFID tag for sensing moisture based on antenna polarization. *Proceedings of the 2014 IEEE International Conference on Electro Information Technology*, IEEE, 542–545, Rapid City, USA.

Sakaki, T., Limsuwat, A., Illangasekare, T.H. (2011). A simple method for calibrating dielectric soil moisture sensors: Laboratory validation in sands. *Vadose Zone Journal*, 10(2), 526–531.

Samineni, V.K., Yoon, J., Crawford, K.E., Jeong, Y.R., McKenzie, K.C., Shin, G., Xie, Z., Sundaram, S.S., Li, Y., Yang, M.Y., Kim, J., Wu, D., Xue, Y., Feng, X., Huang, Y., Mickle, A.D., Banks, A., Ha, J.S., Golden, J.P., Rogers, J.A., Gereau, R.W. (2017). Fully implantable, battery-free wireless optoelectronic devices for spinal optogenetics. *Pain*, 158(11), 2108–2116.

Sample, A.P., Yeager, D.J., Powledge, P.S., Mamishev, A.V., Smith, J.R. (2008). Design of an RFID-based battery-free programmable sensing platform. *IEEE Transactions on Instrumentation and Measurement*, 57(11), 2608–2615.

Sánchez, I., Cortés, M., Riekki, J. (2007). Controlling multimedia players using NFC enabled mobile phones. *Proceedings of the 6th International Conference on Mobile and Ubiquitous Multimedia*, 284, 118–124, Oulu, Finland.

Scanlon, M.G., Roller, R., Mazza, G., Pritchard, M.K. (1994). Computerized video image analysis to quantify color of potato chips. *American Potato Journal*, 71(11), 717–733.

Semacode (2006). Choosing the best 2D barcode format for mobile apps. Technical white paper [Online]. Available at: https://giswiki.hsr.ch/images/d/d2/Best_2D_Barcode.pdf.

Sethia, D., Gupta, D., Mittal, T., Arora, U., Saran, H. (2014). NFC based secure mobile healthcare system. *Proceedings of the 6th International Conference on Communication Systems and Networks*, Bangalore, India.

Shah, S.H. and Yaqoob, I. (2016). A survey: Internet of Things (IOT) technologies, applications and challenges. *Proceedings of the 4th IEEE International Conference on Smart Energy Grid Engineering*, IEEE, 381–385, Oshawa, Canada.

Sheng, L., Dajing, C., Xishan, G., Yuquan, C. (2010). Design of wireless SAW humidity sensor node based on radio wake-up technique. *Proceedings of the 2010 International Conference on Communication Technology*, IEEE, 17–20, Nanjing, China.

Shi, Y., Manco, M., Moyal, D., Huppert, G., Araki, H., Banks, A., Joshi, H., McKenzie, R., Seewald, A., Griffin, G., Sen-Gupta, E., Wright, D., Bastien, P., Valceschini, F., Seité, S., Wright, J.A., Ghaffari, R., Rogers, J., Balooch, G., Pielak, R.M. (2018). Soft, stretchable, epidermal sensor with integrated electronics and photochemistry for measuring personal UV exposures. *PLoS ONE*, 13(1), 1–15.

Shinohara, N. (2012). The wireless power transmission: Inductive coupling, radio wave, and resonance coupling. *Wiley Interdisciplinary Reviews: Energy and Environment*, 1(3), 337–346.

Shinohara, N. (2014). *Wireless Power Transfer via Radiowaves*. ISTE Ltd, London, and John Wiley & Sons, New York.

Sicard, C., Glen, C., Aubie, B., Wallace, D., Jahanshahi-Anbuhi, S., Pennings, K., Daigger, G.T., Pelton, R., Brennan, J.D., Filipe, C.D.M. (2015). Tools for water quality monitoring and mapping using paper-based sensors and cell phones. *Water Research*, 70, 360–369.

Sidén, J., Koptioug, A., Gulliksson, M. (2004). The smart diaper moisture detection system. *2004 IEEE MTT-S International Microwave Symposium Digest (IEEE Cat. No.04CH37535)*, IEEE, 659–662, Fort Worth, USA.

Sidén, J., Skerved, V., Gao, J., Forsström, S., Nilsson, H.-E., Kanter, T., Gulliksson, M. (2011). Home care with NFC sensors and a smart phone. *Proceedings of the 4th International Symposium on Applied Sciences in Biomedical and Communication Technologies*, 1–5, Barcelona, Spain.

Siira, E. and Haikio, J. (2007). Experiences from near-field communication (NFC) in a meal service system. *Proceedings of the 1st Annual RFID Eurasia*, IEEE, 1–6, Istanbul, Turkey.

Silicon Craft (2014a). SIC4310 – RFID/NFC tag IC with UART interface [Online]. Available at: https://sic.co.th/product-docs/factsheet-EN20150402172606-1279123908-sic4310.pdf.

Silicon Craft (2014b). SIC4340 – NFC forum type 2 tag IC with analog sensor interface [Online]. Available at: https://www.sic.co.th/product/sic4340/.

Silicon Craft (2014c). SIC4341 – NFC forum type 2 tag IC with potentiostat sensor interface [Online]. Available at: https://www.sic.co.th/product/sic4341/.

Simik, M.Y.E., Chi, F., Abdelgader, A.M., Saleh, R.S.I. (2014). Automated alarm system for diaper wet using GSM. *Proceedings of the 17th International Conference on Computational Science and Engineering*, IEEE, 1799–1803, Chengdu, China.

Sinclair, D.F. and Williams, J. (1979). Components of variance involved in estimating soil water content and water content change using a neutron moisture meter. *Australian Journal of Soil Research*, 17(2), 237–247.

Singh, G. and Sharma, M. (2015). Barcode technology and its application in libraries and information centers. *International Journal of Next Generation Library and Technologies*, 1(1), 1–8.

ST Microelectronics (2017a). ST25DVxxx – Dynamic NFC/RFID tag IC [Online]. Available at: https://www.st.com/resource/en/datasheet/st25dv04k.pdf.

ST Microelectronics (2017b). M24LR04E-R – Dynamic NFC/RFID tag IC [Online]. Available at: https://www.st.com/resource/en/datasheet/m24lr04e-r.pdf.

Stogryn, A. (1971). Equations for calculating the dielectric constant of saline water. *IEEE Transactions on Microwave Theory and Techniques*, 19(8), 733–736.

Svacina, J. (1992). Analysis of multilayer microstrip lines by a conformal mapping method. *IEEE Transactions on Microwave Theory and Techniques*, 40(4), 4–7.

Syahrir, W.M., Suryanti, A., Connsynn, C. (2009). Color grading in tomato maturity estimator using image processing technique. *Proceedings of the 2nd IEEE International Conference on Computer Science and Information Technology*, IEEE, 276–280, Beijing, China.

Tankiewicz, S., Schaefer, J., Dehennis, A. (2013). A co-planar, near field communication telemetry link for a fully-implantable glucose sensor using high permeability ferrites. *Proceedings of IEEE Sensors 2013*, Baltimore, USA.

TAOS (2012). TCS34725 – Color light to digital converter with IR filter [Online]. Available at: https://cdn-shop.adafruit.com/datasheets/TCS34725.pdf.

Teichmann, J., Burger, K., Hasche, W., Herrfurth, J., Täschner, G. (2003). One time programming (OTP) with Zener diodes in CMOS processes. *Proceedings of the 33rd Conference on European Solid-State Device Research*, IEEE, 433–436, Estoril, Portugal.

Texas Instruments (2009). LM75x – Digital temperature sensor and thermal watchdog with two-wire interface [Online]. Available at: https://www.ti.com/lit/ds/symlink/lm75b.pdf?ts=1630882678743&ref_url=https%253A%252F%252Fwww.ti.com%252Fproduct%252FLM75B.

Texas Instruments (2014a). RF430FRL15xH NFC ISO 15693 – Sensor transponder [Online]. Available at: https://www.ti.com/lit/ds/symlink/rf430frl153h.pdf?ts=1630929643140&ref_url=https%253A%252F%252Fwww.google.com%252F.

Texas Instruments (2014b). TRF7970A – Multiprotocol fully integrated 13.56-MHz RFID and near field communication (NFC) transceiver IC [Online]. Available at: https://www.ti.com/lit/ds/slos743m/slos743m.pdf?ts=1630919811380&ref_url=https%253A%252F%252Fwww.google.com%252F.

Texas Instruments (2018). MSP430F5310, MSP430F530x – Mixed-signal microcontrollers [Online]. Available at: https://www.ti.com/lit/ds/symlink/msp430f5310.pdf?ts=1630903519509&ref_url=https%253A%252F%252Fwww.ti.com%252Fproduct%252FMSP430F5310.

Trambadia, S. and Mayatra, H. (2017). Food detection on plate based on the HSV color model. *Proceedings of the 2016 Online International Conference on Green Engineering and Technologies*, Coimbatore, India.

Van Hoeck, K., Bael, A., Lax, H., Hirche, H., Van Dessel, E., Van Renthergem, D., van Gool, J.D. (2007). Urine output rate and maximum volume voided in school-age children with and without nocturnal enuresis. *Journal of Pediatrics*, 151(6), 575–580.

Vandekinderen, I., Van Camp, J., Devlieghere, F., Veramme, K., Denon, Q., Ragaert, P., De Meulenaer, B. (2008). Effect of decontamination agents on the microbial population, sensorial quality, and nutrient content of grated carrots (*Daucus carota* L.). *Journal of Agricultural and Food Chemistry*, 56(14), 5723–5731.

Vicente, J.M., Avila-Navarro, E., Juan, C.G., Garcia, N., Sabater-Navarro, J.M. (2016). Design of a wearable bio-patch for monitoring patient's temperature. *Proceedings of the 38th Annual International Conference of the IEEE Engineering in Medicine and Biology Society*, IEEE, 4792–4795, Orlando, USA.

Vuong, N.K., Chan, S., Lau, C.T. (2010). PH levels classification by color quantization on a camera phone. *Proceedings of the 2010 WRI International Conference on Communications and Mobile Computing*, 1, 448–452, Shenzen, China.

Wai, A.A.P., Fook, V.F.S., Jayachandran, M., Biswas, J., Nugent, C., Mulvenna, M., Lee, J.-E., Kiat, P.Y.L. (2008). Smart wireless continence management system for persons with dementia. *Telemedicine and e-Health*, 14(8), 825–832.

Wai, A.A.P., Fook, V.F.S., Jayachandran, M., Biswas, J., Nugent, C., Mulvenna, M., Daqing, Z., Craig, D., Passmore, P., Lee, J.-E., Kiat, P.Y.L. (2012). Towards developing effective Continence Management through wetness alert diaper: Experiences, lessons learned, challenges and future directions. *Proceedings of the 4th International Conference on Pervasive Computing Technologies for Healthcare*. Munich, Germany.

Want, R. (2006). An introduction to RFID technology. *IEEE Pervasive Computing*, 5(1), 25–33.

Wei, X., Wang, Z., Dai, H. (2014). A critical review of wireless power transfer via strongly coupled magnetic resonances. *Energies*, 7(7), 4316–4341.

Weinstein, R. (2005). RFID: A technical overview and its application to the enterprise. *IT Professional*, 7(3), 27–33.

WHO (1998). 50 facts: Global health situation and trends 1955–2025. *World Health Organization* [Online]. Available at: https://web.uri.edu/mind/files/CPL-300-HEALTH-WHO-50-Facts-copy.doc.

Wickström, H., Broos, A., Nyman, J.O., Kortesmäki, E., Eklund, P., de Beer, T., Preis, M., Sandler, N. (2018). Handheld colorimeter as quality control tool for inkjet printed flexible levothyroxine doses for pediatric use. *International Journal of Pharmaceutics*, 536(2), 508–509.

Wikner, J.J., Zötterman, J., Jalili, A., Farnebo, S. (2017). Aiming for the cloud – A study of implanted battery-free temperature sensors using NFC. *Proceedings of the 2016 International Symposium on Integrated Circuits*, 1–4, Singapore.

Wisehart, M.K. (1921). Making your imagination work for you. *Tesla Universe* [Online]. Available at: https://teslauniverse.com/nikola-tesla/articles/making-your-imagination-work-you.

WO2013115147A1 (2013). Antenna apparatus and communication terminal apparatus. *Google Patents* [Online]. Available at: https://patents.google.com/patent/WO2013115147A1/en.

Wobschall, D. (1978). A frequency shift dielectric soil moisture sensor. *IEEE Transactions on Geoscience Electronics*, 16(2), 112–118.

Wu, J., Qi, L., Kumar, R.S.S., Kumar, N., Tague, P. (2012). S-SPAN: Secure smart posters in Android using NFC. *Proceedings of the 2012 IEEE International Symposium on a World of Wireless, Mobile and Multimedia Networks*, 1–3, San Francisco, USA.

Wu, C.C., Kuo, W.C., Wang, H.J., Huang, Y.C., Chen, Y.H., Chou, Y.Y., Yu, S.A., Lu, S.S. (2015). A pliable and batteryless real-time ECG monitoring system-in-a-patch. *Proceedings of the 2015 International Symposium on VLSI Design, Automation and Test*, 4–7, Hsinchu, Taiwan.

Xu, K., Sheng, Q., Zhang, X., Li, P., Chen, S. (2015). Design and calibration of the unilateral sensitive soil moisture sensor. *IEEE Sensors Journal*, 15(8), 4587–4594.

Yambem, L., Yapici, M.K., Zou, J. (2008). A new wireless sensor system for smart diapers. *IEEE Sensors Journal*, 8(3), 238–239.

Yu, W., Seo, W., Tan, T., Jung, B., Ziaie, B. (2016). A diaper-embedded disposable nitrite sensor with integrated on-board urine-activated battery for UTI screening. *Proceedings of the 38th Annual International Conference of the IEEE Engineering in Medicine and Biology Society*, 303–306, Orlando, USA.

Yunseop, K. and William, M.I. (2015). Remote sensing and control of an irrigation system using a distributed wireless sensor network. *IEEE Transactions on Instrumentation and Measurement*, 57(7), 1379–1387.

Zargham, M. and Gulak, P.G. (2012). Maximum achievable efficiency in near-field coupled power-transfer systems. *IEEE Transactions on Biomedical Circuits and Systems*, 6(3), 228–245.

Zhao, A. and Chen, H. (2016). Small size NFC antenna with high performance. *Proceedings of the 2016 IEEE International Symposium on Antennas and Propagation*, IEEE, 3, 1469–1470, Fajardo, Puerto Rico.

Zhao, Y., Wang, D., Qian, D. (2009). Machine vision based image analysis for the estimation of pear external quality. *Proceedings of the Second International Conference on Intelligent Computation Technology and Automation*, 629–632, Changsha, China.

Zhao, Y., Smith, J.R., Sample, A. (2015). NFC-WISP: A sensing and computationally enhanced near-field RFID platform. *Proceedings of the 2015 IEEE International Conference on RFID*, 174–181, San Diego, USA.

Zhu, Z. (2004). RFID analog front end design tutorial. Auto-ID Labs, University of Adelaide [Online]. Available at: https://www.iczhiku.com/resourceDetail/AgS6yYgDZoctopdhIdrNdQ==.

Zhu, J.Q., Ban, Y.L., Sim, C.Y.D., Wu, G. (2017). NFC antenna with nonuniform meandering line and partial coverage ferrite sheet for metal cover smartphone applications. *IEEE Transactions on Antennas and Propagation*, 65(6), 2827–2835.

Ziai, M.A. and Batchelor, J.C. (2015). Smart radio-frequency identification tag for diaper moisture detection. *Healthcare Technology Letters*, 2(1), 18–21.

Zierhofer, C.M. and Hochmair, E.S. (1996). Geometric approach for coupling enhancement of magnetically coupled coils. *IEEE Transactions on Biomedical Engineering*, 43(7), 708–714.

Index

L, M

N, P

Q, R

S, T

V, W

Other titles from

iSTE

in

Information Systems, Web and Pervasive Computing

2022

KARAM Elie
General Contractor Business Model for Smart Cities: Fundamentals and Techniques

2021

BEN REBAH Hassen, BOUKTHIR Hafedh, CHÉDEBOIS Antoine
Website Design and Development with HTML5 and CSS3

EL ASSAD Safwan, BARBA Dominique
Digital Communications 1: Fundamentals and Techniques
Digital Communications 2: Directed and Practical Work

GAUDIN Thierry, MAUREL Marie-Christine, POMEROL Jean-Charles
Chance, Calculation and Life

LAURENT Sébastien-Yves
Conflicts, Crimes and Regulations in Cyberspace
(Cybersecurity Set – Volume 2)

LE DEUFF Olivier
Hyperdocumentation (Intellectual Technologies Set – Volume 9)

PÉLISSIER Maud
Cultural Commons in the Digital Ecosystem
(Intellectual Technologies Set – Volume 8)

2020

CLIQUET Gérard, with the collaboration of BARAY Jérôme
Location-Based Marketing: Geomarketing and Geolocation

DE FRÉMINVILLE Marie
Cybersecurity and Decision Makers: Data Security and Digital Trust

GEORGE Éric
Digitalization of Society and Socio-political Issues 2: Digital, Information and Research

HELALI Saida
Systems and Network Infrastructure Integration

LOISEAU Hugo, VENTRE Daniel, ADEN Hartmut
Cybersecurity in Humanities and Social Sciences: A Research Methods Approach (Cybersecurity Set – Volume 1)

SEDKAOUI Soraya, KHELFAOUI Mounia
Sharing Economy and Big Data Analytics

SCHMITT Églantine
Big Data: An Art of Decision Making
(Intellectual Technologies Set – Volume 7)

2019

ALBAN Daniel, EYNAUD Philippe, MALAURENT Julien, RICHET Jean-Loup, VITARI Claudio
Information Systems Management: Governance, Urbanization and Alignment

AUGEY Dominique, with the collaboration of ALCARAZ Marina
Digital Information Ecosystems: Smart Press

BATTON-HUBERT Mireille, DESJARDIN Eric, PINET François
Geographic Data Imperfection 1: From Theory to Applications

BRIQUET-DUHAZÉ Sophie, TURCOTTE Catherine
From Reading-Writing Research to Practice

BROCHARD Luigi, KAMATH Vinod, CORBALAN Julita, HOLLAND Scott, MITTELBACH Walter, OTT Michael
Energy-Efficient Computing and Data Centers

CHAMOUX Jean-Pierre
The Digital Era 2: Political Economy Revisited

COCHARD Gérard-Michel
Introduction to Stochastic Processes and Simulation

DUONG Véronique
SEO Management: Methods and Techniques to Achieve Success

GAUCHEREL Cédric, GOUYON Pierre-Henri, DESSALLES Jean-Louis
Information, The Hidden Side of Life

GEORGE Éric
Digitalization of Society and Socio-political Issues 1: Digital, Communication and Culture

GHLALA Riadh
Analytic SQL in SQL Server 2014/2016

JANIER Mathilde, SAINT-DIZIER Patrick
Argument Mining: Linguistic Foundations

SOURIS Marc
Epidemiology and Geography: Principles, Methods and Tools of Spatial Analysis

TOUNSI Wiem
Cyber-Vigilance and Digital Trust: Cyber Security in the Era of Cloud Computing and IoT

2018

ARDUIN Pierre-Emmanuel
Insider Threats
(Advances in Information Systems Set – Volume 10)

CARMÈS Maryse
Digital Organizations Manufacturing: Scripts, Performativity and Semiopolitics
(Intellectual Technologies Set – Volume 5)

CARRÉ Dominique, VIDAL Geneviève
Hyperconnectivity: Economical, Social and Environmental Challenges
(Computing and Connected Society Set – Volume 3)

CHAMOUX Jean-Pierre
The Digital Era 1: Big Data Stakes

DOUAY Nicolas
Urban Planning in the Digital Age
(Intellectual Technologies Set – Volume 6)

FABRE Renaud, BENSOUSSAN Alain
The Digital Factory for Knowledge: Production and Validation of Scientific Results

GAUDIN Thierry, LACROIX Dominique, MAUREL Marie-Christine, POMEROL Jean-Charles
Life Sciences, Information Sciences

GAYARD Laurent
Darknet: Geopolitics and Uses
(Computing and Connected Society Set – Volume 2)

IAFRATE Fernando
Artificial Intelligence and Big Data: The Birth of a New Intelligence
(Advances in Information Systems Set – Volume 8)

LE DEUFF Olivier
Digital Humanities: History and Development
(Intellectual Technologies Set – Volume 4)

MANDRAN Nadine
Traceable Human Experiment Design Research: Theoretical Model and Practical Guide
(Advances in Information Systems Set – Volume 9)

PIVERT Olivier
NoSQL Data Models: Trends and Challenges

ROCHET Claude
Smart Cities: Reality or Fiction

SALEH Imad, AMMI, Mehdi, SZONIECKY Samuel
Challenges of the Internet of Things: Technology, Use, Ethics
(Digital Tools and Uses Set – Volume 7)

SAUVAGNARGUES Sophie
Decision-making in Crisis Situations: Research and Innovation for Optimal Training

SEDKAOUI Soraya
Data Analytics and Big Data

SZONIECKY Samuel
Ecosystems Knowledge: Modeling and Analysis Method for Information and Communication
(Digital Tools and Uses Set – Volume 6)

2017

BOUHAÏ Nasreddine, SALEH Imad
Internet of Things: Evolutions and Innovations
(Digital Tools and Uses Set – Volume 4)

DUONG Véronique
Baidu SEO: Challenges and Intricacies of Marketing in China

LESAS Anne-Marie, MIRANDA Serge
The Art and Science of NFC Programming
(Intellectual Technologies Set – Volume 3)

LIEM André
Prospective Ergonomics
(Human-Machine Interaction Set – Volume 4)

MARSAULT Xavier
Eco-generative Design for Early Stages of Architecture
(Architecture and Computer Science Set – Volume 1)

REYES-GARCIA Everardo
The Image-Interface: Graphical Supports for Visual Information
(Digital Tools and Uses Set – Volume 3)

REYES-GARCIA Everardo, BOUHAÏ Nasreddine
Designing Interactive Hypermedia Systems
(Digital Tools and Uses Set – Volume 2)

SAÏD Karim, BAHRI KORBI Fadia
Asymmetric Alliances and Information Systems:Issues and Prospects
(Advances in Information Systems Set – Volume 7)

SZONIECKY Samuel, BOUHAÏ Nasreddine
Collective Intelligence and Digital Archives: Towards Knowledge Ecosystems
(Digital Tools and Uses Set – Volume 1)

2016

BEN CHOUIKHA Mona
Organizational Design for Knowledge Management

BERTOLO David
Interactions on Digital Tablets in the Context of 3D Geometry Learning
(Human-Machine Interaction Set – Volume 2)

BOUVARD Patricia, SUZANNE Hervé
Collective Intelligence Development in Business

EL FALLAH SEGHROUCHNI Amal, ISHIKAWA Fuyuki, HÉRAULT Laurent, TOKUDA Hideyuki
Enablers for Smart Cities

FABRE Renaud, in collaboration with MESSERSCHMIDT-MARIET Quentin, HOLVOET Margot
New Challenges for Knowledge

GAUDIELLO Ilaria, ZIBETTI Elisabetta
Learning Robotics, with Robotics, by Robotics
(Human-Machine Interaction Set – Volume 3)

HENROTIN Joseph
The Art of War in the Network Age
(Intellectual Technologies Set – Volume 1)

KITAJIMA Munéo
Memory and Action Selection in Human–Machine Interaction
(Human–Machine Interaction Set – Volume 1)

LAGRAÑA Fernando
E-mail and Behavioral Changes: Uses and Misuses of Electronic Communications

LEIGNEL Jean-Louis, UNGARO Thierry, STAAR Adrien
Digital Transformation
(Advances in Information Systems Set – Volume 6)

NOYER Jean-Max
Transformation of Collective Intelligences
(Intellectual Technologies Set – Volume 2)

VENTRE Daniel
Information Warfare – 2nd edition

VITALIS André
The Uncertain Digital Revolution
(Computing and Connected Society Set – Volume 1)

2015

ARDUIN Pierre-Emmanuel, GRUNDSTEIN Michel, ROSENTHAL-SABROUX Camille
Information and Knowledge System
(Advances in Information Systems Set – Volume 2)

BÉRANGER Jérôme
Medical Information Systems Ethics

BRONNER Gérald
Belief and Misbelief Asymmetry on the Internet

IAFRATE Fernando
From Big Data to Smart Data
(Advances in Information Systems Set – Volume 1)

KRICHEN Saoussen, BEN JOUIDA Sihem
Supply Chain Management and its Applications in Computer Science

NEGRE Elsa
Information and Recommender Systems
(Advances in Information Systems Set – Volume 4)

POMEROL Jean-Charles, EPELBOIN Yves, THOURY Claire
MOOCs

SALLES Maryse
Decision-Making and the Information System
(Advances in Information Systems Set – Volume 3)

SAMARA Tarek
ERP and Information Systems: Integration or Disintegration
(Advances in Information Systems Set – Volume 5)

2014

DINET Jérôme
Information Retrieval in Digital Environments

HÉNO Raphaële, CHANDELIER Laure
3D Modeling of Buildings: Outstanding Sites

KEMBELLEC Gérald, CHARTRON Ghislaine, SALEH Imad
Recommender Systems

MATHIAN Hélène, SANDERS Lena
Spatio-temporal Approaches: Geographic Objects and Change Process

PLANTIN Jean-Christophe
Participatory Mapping

VENTRE Daniel
Chinese Cybersecurity and Defense

2013

BERNIK Igor
Cybercrime and Cyberwarfare

CAPET Philippe, DELAVALLADE Thomas
Information Evaluation

LEBRATY Jean-Fabrice, LOBRE-LEBRATY Katia
Crowdsourcing: One Step Beyond

SALLABERRY Christian
Geographical Information Retrieval in Textual Corpora

2012

BUCHER Bénédicte, LE BER Florence
Innovative Software Development in GIS

GAUSSIER Eric, YVON François
Textual Information Access

STOCKINGER Peter
Audiovisual Archives: Digital Text and Discourse Analysis

VENTRE Daniel
Cyber Conflict

2011

BANOS Arnaud, THÉVENIN Thomas
Geographical Information and Urban Transport Systems

DAUPHINÉ André
Fractal Geography

LEMBERGER Pirmin, MOREL Mederic
Managing Complexity of Information Systems

STOCKINGER Peter
Introduction to Audiovisual Archives

STOCKINGER Peter
Digital Audiovisual Archives

VENTRE Daniel
Cyberwar and Information Warfare

2010

BONNET Pierre
Enterprise Data Governance

BRUNET Roger
Sustainable Geography

CARREGA Pierre
Geographical Information and Climatology

CAUVIN Colette, ESCOBAR Francisco, SERRADJ Aziz
Thematic Cartography – 3-volume series
Thematic Cartography and Transformations – Volume 1
Cartography and the Impact of the Quantitative Revolution – Volume 2
New Approaches in Thematic Cartography – Volume 3

LANGLOIS Patrice
Simulation of Complex Systems in GIS

MATHIS Philippe
Graphs and Networks – 2nd edition

THERIAULT Marius, DES ROSIERS François
Modeling Urban Dynamics

2009

BONNET Pierre, DETAVERNIER Jean-Michel, VAUQUIER Dominique
Sustainable IT Architecture: the Progressive Way of Overhauling Information Systems with SOA

PAPY Fabrice
Information Science

RIVARD François, ABOU HARB Georges, MERET Philippe
The Transverse Information System

ROCHE Stéphane, CARON Claude
Organizational Facets of GIS

2008

BRUGNOT Gérard
Spatial Management of Risks

FINKE Gerd
Operations Research and Networks

GUERMOND Yves
Modeling Process in Geography

KANEVSKI Michael
Advanced Mapping of Environmental Data

MANOUVRIER Bernard, LAURENT Ménard
Application Integration: EAI, B2B, BPM and SOA

PAPY Fabrice
Digital Libraries

2007

DOBESCH Hartwig, DUMOLARD Pierre, DYRAS Izabela
Spatial Interpolation for Climate Data

SANDERS Lena
Models in Spatial Analysis

2006

CLIQUET Gérard
Geomarketing

CORNIOU Jean-Pierre
Looking Back and Going Forward in IT

DEVILLERS Rodolphe, JEANSOULIN Robert
Fundamentals of Spatial Data Quality

Printed and bound by CPI Group (UK) Ltd, Croydon, CR0 4YY
16/09/2026
14963076-0001